10-28-00

To - Rapchaps

best wishes!

Fred R. Graham

WORKING FOR THE CHESSIE SYSTEM

OLDE KING COAL'S PRIME CARRIER

Vandalia Book Company
Box 8392
Huntington, West Virginia 25705

Published by: Vandalia Book Company
Box 8392
Huntington, West Virginia 25705

Memoir
Covers years 1946 - 1980
Library of Congress Catalog Number: 93-61224

ISBN No. 0-961-7545-5-9

CHESAPEAKE and OHIO Lines
SHOWING ROUTES OF
THROUGH SLEEPING CARS
AND CONNECTIONS
ATLANTIC OCEAN

CONTENTS

SIDELIGHT

January 31, 1829

President Jackson,

The canal system of this country is being threatened by the spread of a new form of transportation known as railroads. The federal government must preserve the canals for the following reasons.

One, if boats are supplanted by railroads, serious unemployment will result. Captains, cooks, drivers, hostlers, repairmen and lock tenders will be left without means of livelihood, not to mention the numerous farmers now employed in growing hay for horses.

Two, boat builders would suffer and towline, whip, and harness makers would be left destitute.

Three, canal boats are absolutely essential to the defense of the United States. In the event of the expected trouble with England, the Erie Canal would be the only means by which we could ever move the supplies so vital to waging modern war.

As you may well know, Mr. President, railroad carriages are pulled at the enormous speed of 15 miles per hour by engines which, in addition to endangering life and limb of passengers, roar and snort their way through the countryside, setting fire to crops, scaring the livestock and frightening women and children. The Almighty certainly never intended that people should travel at such breakneck speed.

Sincerely Yours,
Martin Van Buren
Governor of New York

(The letter above was cited in 'No Growth," *The American Spectator,* Jan. 1984.)[92]

The author testifying before the Interstate Commerce
Commission hearing in the B&O control case.

ACKNOWLEDGMENT

This narrative could not have been written without the assistance and counsel supplied by many friends. The author is extremely grateful for their participation and sincerely hopes that this final result reflects honor for their efforts.

My wife, Velma, has demonstrated remarkable forbearance towards the long hours that I have labored on this manuscript which did not allow a lot of time for retirement recreation pursuits.

Some of the incidents recounted in this memoir were only available to a middle management person like myself through information exchanged among employees. All of the events recounted in this narrative occurred from thirteen to forty-seven years ago and time has a way of blurring the memory. These facts should be kept in mind as the reader proceeds.

Two of my former colleagues in the sales end of the coal department have been most helpful in supplying information and suggestions and they have my heart-felt thanks.

The following individuals have been most helpful and cooperative and deserve my gratitude:

Charles H. Moffat
Thomas W. Dixon
Glenn R. Toothman
Stuart McGehee
Howard Skidmore
Tammy Holtzapful
Tom Tonkery
Frances Huddleston

Ted J. Boszec
John W. Rudnicki
Frank Gaddy
J. L. Quarles, Jr.
R. L. Barnett
James R. Lilly
Marilyn Greenhalgh
Bernard A. Dillon

DEDICATION

This book is dedicated to the memory of Charles E. Lawall (1891 - 1973). Charles Elmer Lawall was a talented Christian gentleman born and reared in Pennsylvania who came to West Virginia after WWI to teach at West Virginia University and went to the top as President of the institution. He was widely known and well liked by the national figures in the coal industry from which he received many honors and finished his career as Vice-President in charge of the Coal Department of the C & O Railway Company.

Catasuqua, Pennsylvania, where he was born, is a very small town just north of Allentown and the much larger city of Bethlehem which borders Allentown on the east. He grew up helping his father in his turn-of-the-century drug store.

He was a very good student in school and demonstrated a great curiosity about all things scientific. When it came time to prepare for a career, it was natural that he would enroll in Lehigh University in nearby Bethlehem, the home of the great Bethlehem Steel Company. He graduated with a B.S. degree in mining engineering in 1914.

He went to work as a testing engineer for the Pittsburgh Testing Labs starting on a series of apprenticeship jobs with various companies in the area.

This job hopping period probably resulted from his insatiable curiosity and desire to investigate new pursuits. This varied experience was obviously helpful in his later career as a teacher.

His job at Bethlehem was abruptly interrupted in early 1918 by Uncle Sam who was drafting young men for World War I army duty in Europe. He was back home and discharged in 1919 when he returned to Bethlehem Steel as a research engineer.

On April 29, 1920, while employed at Bethlehem after the war, he married Marjorie Berger whom he had known for a long time. Encouraged by his new wife, he returned to Lehigh to earn his masters degree. He moved to Morgantown and started on a teaching career at the age of 30 in the fall of 1921. A son was born that same year who was also named Charles Elmer, but called "Bud."

The School of Mines was created in 1930 as a separate and co-equal entity in the old Mechanical Hall alongside the College of Engineering. Charles was made the first Director and was well aware of his benefactors and used every opportunity to cultivate and promote the coal industry.

He was appointed Acting President of West Virginia University in 1938, at the age of 47, when a vacancy developed in that office. He advanced to the permanent position in 1939 reflecting the many friends and good impression he had made all over the state during his previous teaching career of eighteen years.

The Vice-President of the C & O Railway Company, Walter Tuohy, sent an emissary to interview Charles and offer him employment as head of the Railroad's Coal Development Office in Huntington where the incumbent, Ira Davis, was past retirement age.

In 1956, the officer in Cleveland who was in charge of the coal department died very suddenly of a heart attack. After a few month's delay, Charles was promoted by the President, Walter Tuohy, to the job.

For all of his career, Charles maintained a keen interest and faithful attendance at the Presbyterian Church and the Kiwanis Club. He was a truly committed Christian gentleman who went out of his way to demonstrate his caring attitude and sincere concern for other people.

The Percy Nichols award, one of the most prestigious national awards of the Society of Mining Engineers, was presented to him in 1962.

Charles was found dead at the age of 82, lying in the kitchen of his apartment due to heart failure in April 1973, ending a life that had been a tremendous influence on the life and times of the coal industry in West Virginia.

☆

INTRODUCTION
(Early History)

The seed was planted for the eventual development of The Chesapeake & Ohio Railway Company in 1780 when the Virginia Assembly in Williamsburg passed an act authorizing the Justices of Greenbrier County to open a wagon road from the Greenbrier Courthouse east across the mountains to the mouth of the Cowpasture River.

In 1784, George Washington, who had liberated the colonies as Commander in Chief of the Continental Army in 1776 and was later to become our first President, made a trip west to the Ohio River to study firsthand "the extension of navigation of the Potomac as well as opening a road from the east to west." Recognizing the importance of binding the middle west to the Atlantic seaboard, he advocated the building of a canal to connect the two sections. As a result of this suggestion, the Assembly of Virginia chartered the James River Company. This led to the incorporation on January 5, 1785, for the purpose of connecting the James and Kanawha Rivers.

The first meeting of this new company convened on August 10 of that same year and elected George Washington Honorary President and authorized a gift of 100 shares of stock for him. Edmund Randolph acted as President until October 5, 1795, when Dr. William Fonsbee, the first Mayor of Richmond, was elected to the office. It built the first canal east along the falls of James River into Richmond, completed in 1795.

In 1812, a committee of which John Marshall, then Chief Justice of the United States was Chairman, began to study the navigability of the James and Kanawha Rivers to the base of the mountains which separated them. In 1820, all rights to the James River Company were taken over by the State of Virginia which extended the canal west 20 miles to Maidens Adventure and built seven and one-half miles through the Blue Ridge Mountains. It also improved the navigation of the Kanawha River in West Virginia from Charleston to Point Pleasant.

The Virginia Assembly chartered the James River and Kanawha Company which on May 5, 1835, was organized and John Marshall appointed Chairman of a Committee to sell stock in the new company. Claudius Crozet, the great French engineer, was also Virginia State Engineer and advocated that a railroad be built to connect the Atlantic Ocean at Hampton Roads with the territory west of the mountains, but his advice was ignored. A canal, 146.5 miles long, paralleling the James River, was built to Lynchburg in 1840 and on to Buchanan in 1851. In 1861, the company was negotiating with a French group to take over the property and extend the canal across the mountains and on to the Ohio River. The estimated cost of this visionary and extremely complicated canal construction from Buchanan to the Ohio River was 26 million dollars, equivalent to several billion dollars today.

By the 13th of February, 1836, the RF&P Railroad had completed its line and operated a train 20 miles north from Richmond to the Rappahannock and Acquia Rivers. Five days later, the Virginia Assembly chartered the Louisa Railroad to build a railroad from the RF&P line west to the vicinity of Taylorsville passing near Louisa Courthouse to the base of the Southwest mountains and possibly on to the town of Harrisburg. The new line reached Gordonsville on April 10, 1840, and on to Shadwell, Virginia, on May 1, 1849, a distance of 65 miles.

On February 15, 1853, the General Assembly of Virginia established the Covington & Ohio Railroad Company and required it to construct a railroad from Covington to a point on the Ohio River north of the Big Sandy River and south of Point Pleasant. This act was a death knell blow to the long proposed canal. The work was started by Charles B. Fisk, an engineer from New York. A total of three million dollars was spent on construction including a bridge over the Guyandotte River in Cabell County, but the action at Fort Sumpter stopped the project.

The Virginia Central Railroad, as the successor to the Louisa was incorporated by the Virginia Assembly and appropriated $100,000 for an extension to connect Hanover Junction to Richmond. The Virginia Central, with the help of a group headed by Claudius Crozet, completed the line west to Staunton on April 1, 1954, and reached Jackson River depot west of Clifton Forge by July 1, 1857. Thereafter, there was a passenger service between Richmond and Jackson River on various schedules. A stage coach line was then operating daily between Goshen, Virginia, and the Ohio River which connected with the railroad. The Civil War put a stop to further improvements and the railroad was in turmoil throughout the conflict.

After the Civil War, the two states of Virginia and West Virginia passed industrial acts acknowledging the consolidation of the Covington and Ohio Railroad with the Virginia Central and to be called The Chesapeake and Ohio Railroad Company. This new company was to take over the project of connecting the Tidewater Port with the Ohio River. On August 31, 1868, a contract was signed by Commissioners representing both states creating the new company.

Major D. H. Whitcomb, who was Engineer and General Superintendent of the Virginia Central during the Civil War as well as a Confederate Officer, became the guiding force in the new company. W. A. Kuper was the main Locating Engineer and demonstrated his skill in locating the line across the mountains. Clairborn R. Mason, an uneducated contractor who had been Captain of Engineers on Jackson's staff during the war was soon regarded as the best in the building of the mountain line. While construction was in progress over the very difficult mountain terrain, financial problems soon arose. After things got so bad that funds were depleted, General William Carter Wickham, who had become President, persuaded C. P. Huntington to take over his post and finance the new road. He had built the first transcontinental railroad when the Central Pacific and Union Pacific were joined at Promontory, Utah, on May 10, 1869.

The city of Huntington owes its name to this railroad baron described in a flyer from Marshall University:

"Huntington's rags-to-riches story would have done credit to one of Horatio Alger's dime novels. Born in Connecticut in 1821, he was one of nine brothers and sisters. His father, a sometime farmer, wasn't a good provider and so the education of young Collis and his siblings was spotty. At the age of 14, he was apprenticed to a neighboring farmer. His wage was $7 a month plus board. Nevertheless, a thrifty Collis saved enough to buy a horse and wagon, along with a stock of merchandise, and established himself, first as a Yankee peddler, and later as a successful owner of a country store."

"In 1848, gold was discovered in California and Collis eagerly joined the Gold Rush. But it was never his intention to pan for gold, instead he correctly sensed there was money to be made in selling picks, shovels, tents and other supplies to the thousands of miners who were crowding into the state. Huntington quickly turned his modest stake into a small fortune, and started looking for somewhere to invest it."

"And so, at age 39, he joined three other California businessmen - Mark Hopkins, Leland Stanford, and Charles Crocker - in financing the construction of a few miles of California railroad. Other successful railroad ventures followed, and railroads would prove to be the driving force for the rest of Huntington's life. He and his partners eventually would become one of the nation's greatest railroad empires, including the giant Central Pacific and later the even larger Southern Pacific."

Huntington had his own dream of a coast-to-coast railroad and envisioned an extension of the C&O on the east to connect with Tidewater and another in the west to reach Cincinnati, the largest city in the United States at that time. He lost no time in making an inspection tour of the entire railroad which required a boat ride down the New River. On the trip he selected a site for the western terminus on the Ohio River which became the city of Huntington.

The last spike in connecting the new 419.3-mile rail line between Huntington and Richmond was driven a little west of Hawk's Nest on January 29, 1873. Major Whitcomb led a special train from Richmond on January 25 which arrived in Huntington on January 29. It returned to Richmond on February 1, pulling several cars of coal from the Kanawha coal fields. The city dwellers welcomed the new source of fuel with great rejoicing. A demijohn of Ohio River water was also brought back and poured into the James with a ceremony celebrating the marriage of the west and the east with rails of steel.

Westward Ho—used by the Virginia Central from 1861 to 1865 and later used on the Huntington Division. Picture taken at Winifred Junction. A. F. Southworth, engineer, "Barney" Hagen, fireman (standing on left of pilot with tallow pot in hand).

Collis P. Huntington
President 1869 - 1888

Major H. D. Whitcomb
Chief Engineer and General Superintendent Virginia Central Railroad, 1861-65. In 1868 appointed Chief Engineer and General Superintendent of the Chesapeake and Ohio Railroad Company

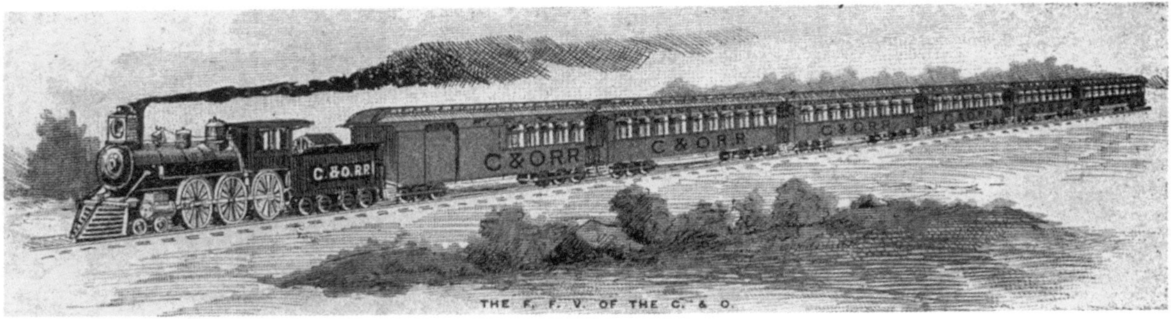

The F. F. V. of the C & O

Coach used by the Virginia Central during Civil War and one of the first passenger coaches used by the Chesapeake & Ohio in West Virginia.

The early years of operation revealed many problems with the difficult terrain and primitive equipment. The coal cars were 20-ton boxes with four wheels. Passenger service between Richmond and Washington in the east and Lexington and Louisville in the west was not very comfortable or dependable. These trains had Pullman cars and stopped for meals at several intermediate points. Passenger trains had air brakes and goose-neck couplers.

Due to financial difficulties, the road went into receivership in October 1875, but ended up as a new company controlled by C. P. Huntington and Associates and the name was changed to Railway. General William C. Wickham again became a receiver when the C&O ran out of funds in October 1887. Drexel, Morgan and Company reorganized the Company in 1888 and installed M. E. Ingalls as the new President succeeding Huntington on October 1, 1888. This event marked the beginning of progressive improvement and expansion that heralded a thriving transportation machine.

Ingalls was a very competent railroad executive who was a lawyer and had extensive experience as President of the Big Four Railroad. He put early emphasis on dependable passenger service and the development of the lucrative coal traffic from the great reserves of high quality coals in southern West Virginia. In May of 1889, the *Fast Flying Virginia* passenger trains were inaugurated. Each train had seven cars painted orange with a red band near the top of the car carrying the name of the railway pulled by a new 10-wheeler engine with 62" driving wheels. The cars were steam heated with electric lights and a diner. That same year the C&O took over the former James River and Kanawha Canal Companies which gave it a low grade rail line along the James River from Clifton Forge to Richmond. The James River and Kanawha Canal Companies had sold in the late 1870's to Richmond and Allegheny Railroad which had built a rail line from Richmond to Clifton Forge by 1881. It went into receivership in 1884. Harry Frazier had been made Chief Engineer and he devised a double-track viaduct in Richmond to connect the James River Division with the Peninsula Division and eliminated the Church Hill Tunnel. It was completed in 1901.

General Manager J. T. Harahan put Harry Frazier in charge of maintenance for the whole system in 1889. He first reconditioned the Cincinnati Division and then extended the work to the rest of the system from his office in Richmond. Numerous improvements were made including better roadbed and ballast, new or repaired bridges and double tracking was initiated. At this time, freight trains were being equipped with air brakes and automatic couplers. One hundred-pound rail was being installed on the main lines.

Mr. Ingalls tried to buy the White Sulphur Hotel and became very frustrated when the owners refused. He then purchased the Hot Springs property and built an extension to serve it in 1891. The C&O Hospital at

M. E. Ingalls
President C. & O. Railway
1888 - 1900

George W. Stevens
President C. & O. Railway
1900 - 1920

General Williams Carter Wickham

W. J. Harahan
President C. & O. Railway
1920 - 1929; 1935 - 1937

Clifton Forge was established on October 27, 1897, and the one in Huntington on May 1, 1900.

Collis Huntington had largely discarded coal as a source of revenue for the railroad and several short extensions were built at Goshen, Longdale and Lowmoor to develop iron ore. Many iron ore furnaces operated for several years, but the business soon dissipated while at the turn of the century coal was becoming the main source of revenue especially from southern West Virginia to Tidewater. Mr. Ingalls began to push New River coals in New England and New York City. This business boomed until over two million tons a year was being sent to New York City displacing a large part of the market for anthracite coal off the Pennsylvania Railroad. To stop the competition, the Pennsy and several other roads bought control of C&O and N&W and raised the freight rate on coal to Tidewater from $1.00 to $1.40.

In order to create a new market for coal, the C&O bought an interest with Sir Christopher Funness and organized the C&O Steamship Company with seven steamships. These ships were later taken over by the British during the Boer War.

Colonel William W. Page lived in Ansted, West Virginia, in a large elegant house and, as part owner, established the Gualey Mountain Coal Company Mine there. He learned a lot about railroads when the C&O built an extension to serve the mine. He promoted a shortline railroad, Deepwater Railway, to serve a lumber mill on Loup Creek and connected it to the C&O at Deepwater. While he was building this line in 1902, a part-owner of the mill, Henry Huddleston Rogers, a multimillionaire from New York City, came to see the mill. Colonel Page convinced him that a railroad could be built south off the Deepwater Railway to serve the large undeveloped coal reserves lying between the C&O and N&W. He also suggested that the rail line could be extended east through Virginia to the coast to profit from the lucrative coal export market. The two men later developed the Tidewater Railway in Virginia to carry out the plan. When the C&O and N&W Railroads realized what was going on, it was too late and after a bitter court battle the Virginian Railway delivered coal in 1907 to a pier in Norfolk by connecting the Deepwater Railway and Tidewater Railway at a total investment of 30 million dollars. The C&O even tried to build a branch line into the contested area to thwart the new competition, but was forced by the local court to turn over a tunnel needed for the Virginian main line south to Mullens. This new railroad proved to be a fierce competitor of C&O for Tidewater coal deliveries and later developed a western connection with the NYC by building a bridge across the C&O main line and the Kanawha River at Deepwater. It was later to play a key role in setting off the big railroad merger movement in the 1960's.

In 1900 the Pennsylvania and New York Central Railroads acquired controlling interest in C&O and replaced Ingalls with George W. Stevens as President. In 1907, NYC pulled out, but Stevens was retained and a new man, M. J. Caples, became Operating Vice-President. Production of coal began to exceed the car supply on the railroad and in 1908 a car allotment commission was established to control the distribution of empty cars to all the coal mines. John W. Davin headed up this effort in the early days and later became President of the Nickel Plate Railroad. In 1920, D'Arcy Morton prepared and published a system of rules and regulations for car supply on the C&O which was copied by most other coal carrying railroads.

Many improvements were made during the Stevens regime. The acquisition of the Hocking Valley Railway made the C&O a north-south line from Toledo, Ohio, to Charleston, South Carolina, by virtue of a connection with the C&O at Elkhorn City, Kentucky. A second track was added to the mainline between Clifton Forge and Cincinnati except for a short stretch along the lower New River and through tunnels in the mountains. Most traveled portions of the main line were beefed up with 130-pound rails and automatic train control. Several branch lines were added in the coal fields including the Piney Creek line into the heart of the low volatile measures in Fayette and Raleigh Counties. Several large coal properties were purchased by Western Pocahontas Corporation and other subsidiary companies of the C&O.

After the turn of the century, the C&O made profits and paid dividends every year except one, 1915. During the same period, many other roads were having great difficulty staying solvent. The advantages of the C&O over others were threefold. The coal fields of southern West Virginia offered the highest quality and most economically recoverable coal resources in the whole world. The C&O was laid out in a very efficient geographic system whereby it has a single route mainline which culminated at both ends (Great Lakes and Atlantic Ocean) where huge volumes of coal were dispatched. Eastbound traffic was served by a large efficient yard at Clifton Forge, Virginia, and westbound traffic at a similar facility in Russell, Kentucky. Six major coal fields with a large variety and type of coal were fed through branchlines to the mainline between the two great yard facilities. The only obstacle that had to be subdued was the mountains separating the two Virginians and good engineering and operating procedures overcame this problem.

On the other hand, only the N&W Railway, the foremost competitor of C&O, enjoyed similar geographic and strategic advantages. Other large eastern railroads were plagued by round-about mainline hauls and diverse destinations and routing. The prime example was the B&O whose coal fields were scattered in such a manner that each required special servicing. Before 1890, the C&O could have built a branch line from Hinton along the

Blue Stone River to reach the great Flat Top coal fields north of Bluefield and thereby usurp the main reason for the development of the N&W. However, on the positive side, the constant threat of competition from the N&W has been a significant factor in prodding the C&O to excel.

The third advantage that contributed to the C&O success was in the experience of its management personnel. Outstanding leadership by people like George Washington, Claudius Crozet, H. D. Whitcomb, George McKendree, Williams Carter Wickham, Collis P. Huntington, M. E. Ingalls, Harry Frazier and George W. Stevens was critical to the success of the railway.

After twenty years on his job, Mr. Stevens died and was replaced as President by W. J. Harahan on December 7, 1920. Harahan was educated at St. John's College in New Orleans and came to the C&O in 1889 as engineer of maintenance after which he served with several other roads including the Seaboard before retiring.

In 1922, a momentous change in the affairs of C&O was inaugurated when the Van Swerigen brothers in Cleveland, Ohio, decided they needed the C&O to match up with the Nickel Plate Railroad and purchased control by acquiring the stock owned by Mrs. H. E. Huntington, the widow of Collis P. Mr. O. P. Van Sweringen became Chairman and Harahan continued as President. The brothers were big-time operators who built the Terminal Tower in Cleveland and moved the C&O headquarters there to the 34, 35, and 36th floors.

By the year 1924, the brothers decided to consolidate four roads, C&O, Nickel Plate, Erie and Pere Marquette, into a single system, but were stymied by the Interstate Commerce Commission. By 1927, the C&O had acquired control of the Erie and Pere Marquette. The I.C.C. permitted C&O to control the PM, but denied the Erie affiliation.

In 1929, the brothers organized the Alleghany Corporation as a holding company for their various interests including the C&O. A new president was elected on May 22, 1929.

J. J. Bernet was reared and educated in New York State. He was a former Trainmaster who had been President of the Nickel Plate and also the Erie Railroad. He inaugurated the *George Washington* passenger train in 1932 with the latest Pullman cars and air conditioning.

Bernet brought with him to the C&O a journalist by the name of L. C. Probert and made him a Vice-President. Probert had been Chief of the Washington, DC, AP news bureau. While in the Capitol, he became well-acquainted with President Wilson. He suggested the name for the train to be called *"Sportsman"* and also the designation for C&O as *George Washington's Railroad* which was printed in large letters near the top of the C&O Building in Huntington. He was asked to devise a new slogan for the passenger service and was considering the phrase *"Sleep like a Top"* when he saw an

J. J. Bernet
President C. & O. Railway
1929 - 1935

R. J. Bowman
President C. & O. Railway
1946 - 1948

George D. Brooke
President C. & O. Railway
1937 - 1942

C. E. Newton
President C. & O. Railway
1942 - 1946

etching by Guido Grunewald in the New York Herald Tribune of a sleeping cat. He obtained permission to use the picture on the C&O and started the slogan: *Sleep like a Kitten* in 1933. The great depression finally spelled the end of the Van Swerengen's business empire and Robert R. Young and associates acquired a large block of Alleghany stock.

Mr. Herbert Fitzpatrick who was General Counsel of the C&O under the Van Sweringens played a major role in transferring their ownership after their financial collapse as related in the Huntington newspaper:

"Mr. Fitzpatrick joined the C&O in Huntington in 1895 as local counsel and moved up through the ranks. He was Board Chairman from 1937 to 1940."

"Mr. Fitzpatrick took over management of the vast Van Sweringen railroad empire in December, 1936, after the deaths of the two bachelor brothers, Mantis J. and Oris P. Van Sweringen, who had assembled the $3,000,000,000 holdings during the years between 1916 and 1929."

"George A. Ball, Muncie, Indiana, glass manufacturer, and George A. Tomlinson, Great Lakes shipowner, who bought a substantial part of the Van Sweringen interests at public auction in 1935, selected Mr. Fitzpatrick to run the post-Depression remains of the empire because of his intimate connection with the two brothers as General Counsel of the Chesapeake and Ohio from 1923 to 1935."

"Oris Van Sweringen died in November, 1936, in Mr. Fitzpatrick's private railroad car while enroute to New York from Cleveland to see representatives of J. P. Morgan & Company. The Morgan office headed a banker's group that had ordered the properties auctioned."

"Messrs. Ball and Tomlinson formed the Mid-America Corporation to operate the major part of the Van Sweringen holdings, and they named Mr. Fitzpatrick its head. Mid-America had acquired control of the Alleghany Corporation which directly or indirectly controlled the C&O, the Erie, Pere Marquette, Missouri Pacific, Chicago and Eastern Illinois, Wheeling and Lake Erie, the New York, Chicago and St. Louis, and other railroad properties."

G. D. Brooke became the new President on December 29, 1937. He was replaced on December 15, 1942, by Carl E. Newton, a New York lawyer, when Young and Associates finally gained control of Alleghany. He was appointed as deputy coal mine administrator in the middle of WWII in 1943, and received the Order of Merit for his work. Robert J. Bowman was President of the PM at the time, and when C&O and Pere Marquette were merged on April 23, 1946, he became the President of the merged company. He was employed by the Pere Marquette as a high school graduate at age 16 and worked up to the top spot.

During the second World War, the C&O was a major link in the great war machine that was developed to win the life or death struggle. Troops, war material and coal were moved with great efficiency and the railway prospered. A. T. Lowmaster and G. J. Geyer were skilled in all phases of the operation and more especially in communication and signals.

Robert R. Young, called the "Gadfly of the Rails" by *Life,* was the prime mover on the C&O from 1942 to 1954. Everything the C&O did in that period was a reflection of Young's personality and his philosophy of what railroading in America ought to be like. He put the C&O at the cutting edge of technology and made it the testing ground of scores of innovative ideas.—*C&O Ry.*

Herbert Fitzpatrick, an important figure.
Photo courtesy of C & O Historical Society

Coal traffic furnished the steady inflow of profits which undergirded the success of the company. After the turn of the century, coal traffic continued to increase in volume toward a peak in 1944 as shown below:

YEAR	MINES	VOLUME (Tons)
1902	105	4,164,260
1905		8,523,300
1915		28,215,450
1925	181	56,086,275
1944		62,805,970
1945	280	56,917,650

Robert R. Young was a dynamic financial man who aspired to become the greatest Railroad Baron of all time. He was convinced that railroad passenger traffic was going to grow by leaps and bounds after the war and he used the profits generated by C&O coal traffic to finance a new era in passenger service. He envisioned a transcontinental railroad in the near future, an all new luxury train to operate between Washington and Cincinnati. He had the C&O to order the streamlined stainless steel equipment for the *Chessie* to be powered by a coal-fired steam turbine locomotive. He ran full page advertisements in several leading newspapers deriding the fact that a hog could cross the country without changing trains but a passenger couldn't.

Source:

Chessie's Railroad," A History of The Chesapeake and Ohio Railway, by George Selden Wallace, 1946?

"Recollections," Harry Frazier, Chief Engineer of C & O, published by C & O Public Relations Department, 1938.

"The Virginian Railway," H. Reid, Kalmback Publishing Company, Milwaukee, Wisconsin, 1961.

"The Chesapeake and Ohio Railway," Patrick C. Dorin, Superior Publishing Company, Burbank, California, 1981.

"Chessie's Road," C. W. Turner, Thomas W. Dixon, Jr., E. L. Huddleston, The C & O Historical Society, Revised Second Edition, 1993.

☆

PREFACE

Modern Railroads, a national magazine out of Chicago, ran a series of articles on "Private Enterprise at Work." One of these articles featured the C&O in November, 1954, and consisted of 73 pages. It is excerpted here to give the reader a background appreciation for the high esteem in which the Railway was generally recognized during the post WWII period:

"Throughout the 5100-mile Chesapeake and Ohio Railway System -- and integrated with a vast $460 million postwar improvement program -- a new dimension is being added to railroad activity."

* * * * *

"Its workings are evident in many places on many other railroads; but nowhere as extensively as on the C&O. It cannot be defined as a single type of approach; but rather runs the gamut of present day corporate responsibility -- from customer relations to physical and operational modernization, to the very environment in which human beings do their work."

* * * * *

"To implement this policy of orderly, scientific cost control and advanced planning, C&O's Finance Department has been converted from a purely accounting function to a service organization. Its primary duty now is to supply Traffic, Engineering and Maintenance Departments with up-to-the-minute cost reports, tailored to their individual needs, plus figures and forecasts that they can use in making decisions and planning their budgets."

* * * * *

"The advanced management approach of the C&O, directed by Walter J. Tuohy, President, and with the complete support of Cyrus S. Eaton, its Chairman of the Board, has vitalized the corporate imagination. It has nurtured an inclination toward looking at basic railroad problems and policies in terms of long-range profit and loss characteristics. And it has encouraged the railroad to look to other industry, as well as the railroad industry, to make sure that no opportunity is lost for bringing about a better working, more efficient railroad service organization."

"Dramatic evidence of this approach is C&O's recently signed contract with Remington Road, Inc., for the lease of a Univac computer system. The 'electronic brain,' as it is more popularly known, was generally considered to be in the distant future as far as its practical application to railroad work goes. But C&O staff experts made intensive studies of its use in other industries and were quick to recognize its practicality in simplifying complicated railroad paper-work procedures. C&O has been the first railroad to order a general purpose computer for such a revolutionary application."

"Like other progressive businesses, C&O also recognizes the value of cultivating the good will of the public. Its organized public relations activity is extensive. But, more important, 'public relations mindedness' is ingrained among C&O officers and personnel. This is well illustrated in C&O's staff meetings, where the public relations aspects of all decisions are carefully weighed."

"Tuohy's staff officers are all men of outstanding ability and strong personalities; but through even-tempered skillful leadership, Tuohy has welded them into a smooth-running, harmonious management group. Keenly sensitive to human reactions, he takes great pains to make sure that all officers understand that the door to his office is open; that no one ever has cause to feel that he or his problems have been brushed aside."

"However, Tuohy dislikes making decisions for or between his officers; instead he tries to direct their thinking so that they will solve their own problems. 'Business today is far above the capabilities of one man,' Mr. Tuohy comments. 'You must have a management team, men who are devoted and loyal to the property, themselves and to the people who work for them."

"This management team is the number one factor in the success of any business organization, Mr. Tuohy believes. 'Its members must be men of integrity, hard-hitting and particularly in the railroad business -- on their toes, progressive and hard-working."

"That is especially necessary in the railroad business because of the many regulations and restriction that surround it. After hitting their heads against the wall for years, railroad men sometimes come to accept conditions. We must not get into that state but make the best of conditions and improve in the areas where we can still move."

"The railroads must tell their story honestly, persuasively and unselfishly. And one of the best ways to get the public and the government on their side is to improve their services, which have not always been what they should be."

* * * * *

"Of these activities, improving the quality of supervision through education and training is perhaps the most important. Work simplification, which C&O terms its 'inside efficiency engineering,' is now being used in several C&O departments. It is a means of cutting costs, but C&O has found a main benefit to be improved morale and better human relations."

"C&O also believes in the incentive of stock ownership by both officers and employees. Its stock option plan for officers now applies to 75 officers. More than two thousand employees have purchased stock under the road's unusual employee stock-purchase plan."

* * * * *

"Almost overshadowing its dramatic advanced management approach (because the effects can be seen more readily) is the vast improvement program which C&O embarked upon at the end of World War II."

"The program is evident in huge fleets of diesels, new fleets of freight and passenger cars, multi-million dollar terminals, new roadway equipment, modernized shops, centralized traffic control, expanded communications and new trainferries."

"The program has totaled $460 million rising to a peak of $80 million in 1952 alone. It is now completed. Capital expenditures dropped to $49 million last year and will be much less this year. Because it had the providence and foresight to make its improvements while the money was available, C&O is better able today to cope with today's reduced traffic level."

* * * * *

"With 5100 road miles, the Chesapeake and Ohio Railway is 17th in size among the railroad systems of the United States. Yet last year it ranked 7th in gross revenue and 4th in net income."

"But fortunate location is only one part of the C&O story. To an even greater extent, the improved earnings of recent years stem from the policies of its progressive management. The modern C&O is vigorously going after new and diversified business and is applying advanced techniques of cost control to its internal operations."

"Almost three-quarters of the $460 million improvement program has been financed internally -- primarily via depreciation and accruals and retained earnings. Much of the additional money required has been obtained in the form of low-cost equipment obligations."

* * * * *

"What C&O has done to modernize that plant; and what it plans to do in the future with aid of the management concepts it is developing and putting into effect, are described in the articles that follow."

"C&O's management's approach to developing a better railroad team includes extra benefits which all employees can share equally."

"When speaking of these 'plus' features, many C&Oers will more than likely put special emphasis upon their unique Employees' Stock Purchase Plan, inaugurated in 1952."

* * * * *

"Through the first of September, 1954, C&O employees had purchased over 35,000 shares under the plan. An employee who signed up when the plan started in 1952 would have paid $270 in payroll deductions -- but he would also have been credited with $67.50 in dividends."

* * * * *

"Another popular extra is the company-paid supplementary retirement pension. The Supplementary Retirement Plan is funded and is non-contributory for employees with salaries over $350 per month. Its benefits supplement those of the Railroad Retirement Act."

* * * * *

"Another big extra-curricular activity of C&O employees are system-wide Better Service Conferences. These meetings often are spiced with entertainment and good fellowship, but they also have their more serious side."

* * * * *

"Better Service Conferences spread to two additional cities and today they total 26, in as many important C&O communities all the way from Newport News, Virginia, to St. Thomas, Ontario."

"Last year about 10,000 employees attended one or more of the Better Service Conferences. And about 250 participated actively as officers, conference leaders and committee chairmen, devoting considerable of their own time to conference activities. Officers of each conference are elected by employees active in that conference."

* * * * *

"Tracks Magazine now carries a monthly feature, Better Service Notes. Bulletin boards also feature Better Service with announcements and posters. They also highlight the importance of courtesy and friendliness to better service to the public."

* * * * *

"Far reaching changes have taken place in C&O's fiscal and accounting activities since 1949, when progressive John E. Kusik became Accounting Vice

President. (He's now Vice President-Finance with jurisdiction over accounting, treasury and tax work.)"

* * * * *

"Essentially, the C&O Finance Departments consist of three producing, or line, departments -- Treasury, Tax, and Accounting -- plus two staff groups, or 'arms' of the Vice President."

"The Treasury and Accounting Departments perform basically the same functions as do similar departments elsewhere, with the notable exception that, as both Comptroller Roger Brown and Treasurer Joseph Anzalone sum it up: 'We're action-oriented now, instead of history-oriented.' And that has come to mean serving other departments -- Traffic, Transportation, Maintenance, as well as top management -- with prompt, accurate facts in a way that will stimulate the management to plan constructively for the future."

"Additionally, the Comptroller has his on his staff a qualified group of internal auditors to maintain a continuing check on the adequacy of its accounting work."

* * * * *

"All proposed major improvements in the department are first 'screened' through the Planning Group, after which they are designated by an MCO (Major Current Objective) number and required line and staff manpower is assigned. Usually such changes require the assistance of Management Services staff personnel and are of a long-term nature. The Planning Group also decides on the priority."

* * * * *

"Evaluation is the key feature of the entire personnel program; for to overrate or underrate an individual would be harmful both to the man and to the company. C&O now uses a detailed 'Management Evaluation and Development Guide' which each evaluator fills out. This guide includes questions about the candidate's ability to make decisions,*** his organizing ability, responsibility, adaptability, job knowledge and general knowledge and understanding."

"Next comes training. A development plan is 'tailored' to fit each candidate and may include either informal or formal courses outside of those run by the Finance Departments."

* * * * *

"Just about the most dramatic aspect of C&O's advanced management approach is how quickly vital statistics are compiled and the useful form in which they are presented to the railroad's various management."

* * * * *

"Reports prepared for the top management come out within a very few days after the close of the month they cover. They are concise, with the data 'predigested' and presented in a form that helps busy officers to whom the reports are directed to get the most out of them quickly."

* * * * *

"By noon of each following day, a report is distributed to each member of management, providing them with a good estimate of the day's revenues."

* * * * *

"Monthly figures also are now processed quickly. A preliminary report, highlighting the results of the month's operations, is distributed by the fourth working day after the close of each month to the Board of Directors, department heads and other key officers."

* * * * *

"Centralization -- and almost complete mechanization -- of Car Accounting on the Chesapeake and Ohio was accomplished well before the recent formal merging of its Pere Marquette and Chesapeake Districts."

"The central Car Accounting Office is now located at C&O's operating headquarters in Richmond, Virginia. In planning the new central Accounting Office, C&O recognized that tape-to-card or even more advanced data transmission techniques may one day have to be integrated into the system. In fact, the entire office was planned on work simplification principles so that work progresses smoothly through all its stages with a minimum of lost motion."

* * * * *

"Dramatic evidence of C&O's advanced management approach is the railroad's recently signed contract with Remington Rand Inc., for the lease of a Univac computer system, popularly known as an 'electronic brain.' The contract followed intensive studies by Finance Department methods groups aimed at simplifying complicated paperwork procedures."

* * * * *

"The Finance Department study was made by its own research office under A. F. Dell Isola and J. F. Feagler. Their findings on the effects of computerizing C&O's varied and complex paperwork procedures were so far-reaching as to justify lease of an electronic computer."

"Meanwhile, the study of long-range applications of a computer is continuing. This study so far has indicated that electronic data processing holds great promise of achieving economy and greater efficiency through integrating the information requirements of several departments. Such data processing will have a tremendous impact on the railroad's communications system. As a result, a new Teletype system is being installed on C&O's main lines designed for possible future integration with a centralized electronic data processing operation."

* * * * *

"Last year, Engineering officers also became interested in the possibility of using the Operations Research approach to analyze the factors governing rail replacement in the main lines, and to come up, if possible, with a scientific formula for determining the most economic rail renewal practices for the conditions prevailing on the Chesapeake and Ohio."

* * * * *

"The data is now being correlated, and a report of the findings is expected in the near future. C&O feels that carefully analyzed results of the investigation can help in forecasting budget expenditures for rail by enabling estimate to be made of new rail requirements for the assumed traffic levels."

"Still another management aid of C&O's Finance Department that goes far beyond traditional railroad practice is its cost control work."

"Three years ago an Office of Cost Research was set up to assist the Operating and Maintenance officers in developing effective methods of cost control. To date, most Cost Research work has been done for, and in cooperation with, the Mechanical Department. Ultimately the Office expects to extend its assistance to Transportation and Maintenance of Way officers as well."

* * * * *

"The 'Line Functions' of C&O's Finance Departments also reflect the C&O's advanced management approach. In both Traffic and Operations accounting, mechanization and the streamlining of procedures have been developed to a high degree."

* * * * *

"About 80 percent of the carload freight data is processed mechanically. That includes interline received waybill abstracts, recapitulations and summaries of abstracts, and recapitulations of intermediate proportions. The other 20 percent consists of non-repetitive or other shipments that do not lend themselves to the preparation of master cards."

* * * * *

"Mechanized payroll procedures include the use of electronic calculators which process 6000 cards an hour. Master cards for every employee are used to prepare the time sheets and time cards sent out to work locations. They're also used to produce detail cards into which the hours and days are punched from the returned time sheets. The completed detail cards are then run through an IBM 407 to produce payroll registers, pay checks, and deduction notice cards."

* * * * *

"Progress on the C&O depends upon its people,' emphasizes President Walter Tuohy in discussing C&O's approach to railroad management."

"You can have the finest physical plant in the world,' he continues, 'but without the right kind of personnel you will get nowhere."

* * * * *

"Courtesy Booklet for conductors entitled 'Mr. C&O' was so successful that another is being made for all employees who meet public."

* * * * *

"The application of advanced personnel practices to railroad activity gets its greatest challenge in bringing out the latent abilities of railroaders. C&O's management is using many techniques to meet that challenge effectively. It has applied them also with the definite intent of strengthening the framework or railroading."

"The supervisor, its first line of management, became the key to its improved human relations program. That's because the C&O believes that its supervisors are the real personnel managers of the railroad. Supervisors fill the major role in successfully hiring people, developing them, establishing good work habits, creating a friendly atmosphere, communicating company information and policies, and exemplifying the company."

* * * * *

"A short training course also has been given to 5000 employees in the effective and courteous use of the telephone. Like many C&O training programs, it utilized the conference technique with the participants actually practicing correct telephone methods."

* * * * *

"In 1947, the handling of grievances was a particularly tough thorn in the side of C&O Labor Relations. Since then, the atmosphere has cleared remarkably."

* * * * *

"Explains one official in C&O's Labor Relations Department, 'We have restored to the local officer that badge of office which should have been his all along -- the authority to handle locally and quickly those grievances for which a policy has already been determined."

* * * * *

New River Canyon, Looking East From Hawk's Nest

"There are other benefits as well. Most important is the big boost in morale the system has given to both the local supervisors and the men under them."

* * * * *

"C&O shows its recognition of the importance of the public relations staff function in having the director of public relations report directly to President Tuohy."

* * * * *

"C&O's Annual Report makes liberal use of photographs, picture stories, modern layout and typography and clear, concise writing."

* * * * *

"Symbolizing the friendly spirit of C&O is its famous "Chessie," which has appeared in advertisements and on calendars and playing cards for twenty-one years. Starting in 1948, the public relations department merchandising promotion section has arranged with leading manufacturers to bring out a line of Chessie merchandise, which gives C&O's mascot even wider distribution without cost to the railroad."

* * * * *

"Another unique phase of C&O's public relations program is the Can-Do Special. This is a full size copy of the cab of a C&O 490 Series locomotive which is placed in leading department stores throughout the country to give young people the thrill of 'driving' a real locomotive."

* * * * *

"C&O's dynamic advanced management approach is reflected in its advertising."

* * * * *

"Take the way it is getting across the point of the importance of the traffic manager in modern business."

* * * * *

"The advertisements appear in general business publications and explain that C&O sponsors the campaign 'in the belief that a better understanding of the Traffic Manager's job will contribute to the better and more economical movement of material."

* * * * *

"Thousands of congratulatory letters about this campaign have been received from top management and industrial traffic men. 'Time' magazine prepared a mailing piece using the campaign as a study in successful advertising."

* * * * *

"Still another important C&O advertising program spotlights coal, the 'Fuel of the Future!' And that, too, hasn't been seen by many railroaders because it runs in such publications as Industry Heating, Plant Engineering, Southern Power & Industry, and American School Board Journal."

* * * * *

"When it comes to passenger service, C&O knows that its own territory represents its best potential passenger market and concentrates most of its advertising efforts in those cities it serves directly. Newspaper advertising is supplemented by direct mail, travel literature, posters, window displays and radio."

* * * * *

"It was in connection with nationwide promotion of the new George Washington train in the early 1930's that Chessie, C&O's renowned sleeping kitten, first gained fame as a C&O symbol."

* * * * *

"To date, more than four million calendars featuring Chessie and her family have been distributed to C&O friends all over the world. More recently, she's appeared on various merchandise items."

* * * * *

"For many years the Chesapeake and Ohio has been America's leading hauler of bituminous coal. Last year more than 56 million tons of it were loaded at mines served by the C&O -- equivalent to just about one-eighth of the total national production. Including overhead tonnage and that received from connections and terminated, the C&O during 1953 carried close to 70 million tons of bituminous coal."

* * * * *

"In the coal fields of Kentucky and West Virginia, new mines are constantly being opened and old ones closed. Just since 1946, the C&O has constructed, either by itself or jointly with other railroads, 23 branch line extensions or spurs to tap new mining property. Such construction has aggregated 110.69 miles and cost some $32 million."

"C&O's modern coal terminals at Newport News, Virginia and Toledo, Ohio, are essential elements in its mine-to-consumer transportation system."

* * * * *

"Today the Development Bureau, headed by Assistant Vice President Dr. C. E. Lawall, covers a much wider range of activities. Working closely with mine owners and operators, it serves as liaison between them and the railroad. 'For example,' explains Dr. Lawall, 'when a new mine is opened and we're asked to build a spur or branch to it, we must know what the chances are of that mine's succeeding and how long it is likely to produce. And when a mine is depleted, we must be in a position to recommend a new property for its owner to develop. In short, we must know everything possible about coal mines and mine properties -- inside and out."

"Since 1946, 82 new coal mines have been operated on the C&O. Although 35 of them were not operating in 1953, the remaining 47 produced about 9 million tons -- or 17 percent of the C&O's total coal loadings for that year."

* * * * *

"Today's much broader base for the C&O's freight revenues is the result of farsighted planning on the part of its management. Right after World War II, it undertook to diversify by building up merchandise revenues, while still retaining leadership in coal. Improved service and a campaign to attract new industries to its rails become important parts of this long-range program."

* * * * *

"This varied merchandise traffic moves in scheduled freight trains; and the recent improvements in the manifest train schedules are yet another reason why C&O merchandise traffic is booming. All principal and many secondary lines are served by high-speed, closely-timed schedules."

"C&O's most famous trains are, of course, No. 91, the Speedwest and its eastbound counterpart, the Expeditor, No. 90. They provide top-quality transportation along the main line between Newport News and Louisville, Chicago, Columbus, Toledo, Buffalo, and points on the west shore of Lake Michigan."

* * * * *

"Capitalizing on the full potential of the Lake Michigan trainferry service, incidentally has been one of the cardinal points in C&O's planning for bigger merchandise revenue. The three lake Michigan trainferry routes form a direct bridge between the Northwest and East along which fruit and forest products move eastward and merchandise moves westward. The ferries carry freight cars, automobiles, and passengers between Ludington on the east bank of Lake Michigan and Kewaunee, Manitowoc, and Milwaukee on the west shore."

* * * * *

"A new note in the romance of C&O railroading has been added by the smooth flow of coal and merchandise trains under the certain control of CTC . . . by enlarged and mechanized classification yards . . . by increasing use of radio to converse directly about operating problems as they occur."

"Improvements such as these make present day C&O railroading a far cry from that of only a few years ago. And still to come are more dynamic innovations such as C&O's hybrid box-car, roadrail trailer development."

* * * * *

"Back in the mine country, short groups of cars are combined on the mine branches. Then mine run trains are consolidated into longer transfer runs at the junctions of the large subdivisions and branches. As a result, long, heavy trains reach the C&O main line, badly in need of classification. To do this massive job efficiently, the C&O has three modern retarder-equipped hum yards -- Russell and Stevens, Kentucky and Walbridge, Ohio. Newport News at the Atlantic terminus and Clifton Forge Yards utilize humps but require car riders."

* * * * *

"Out of the North came the relentless, rumbling diesel to Chesapeake and Ohio rails. Up in former Pere Marquette territory, the C&O gained its first experience with diesel operation. Here passenger and freight diesels as well as switchers had established new levels of economical operation for some year prior to 1950. It was in September of that year that the Southern Region placed its first road freight diesel in operation -- on the Chicago Division. Almost at once the advantages of dieselization became clearly apparent, and C&O management began providing diesels in a steady stream."

* * * * *

"In approximately eight years from the time that the system received its first diesel, the C&O has become virtually dieselized -- a remarkable achievement for a railroad whose roots reach into the heart of the coal country. Today only some 100 steam locomotives remain in service, and these are largely assigned to the coal fields of West Virginia and Kentucky. They operate in mine run service and assembly yards, being used constantly while the mines are in operation. At other times, the road is fully dieselized."

* * * * *

"Why, it has been asked, should the C&O, with passenger service only a small portion of its entire business, be so insistent about emphasizing the quality of that service? Here, in the words of Mr. Tuohy, is part of the answer: 'Passengers are the railroads strongest potential support in helping them to get out from under strangling state and federal regulation. They must be given good service.' Also, C&O realizes the public relations value of good passenger service and the effect it can have upon the shipping world's opinion of a railroad as an efficient freight carrier. It also recognizes the responsibility it has to the travelers and businesses in the substantial communities along its line -- freight traffic sources all."

* * * * *

"As a result of these efforts to reduce deficits, C&O currently is operating almost three million less passenger train miles annually than in 1948. Its out-of-pocket deficit was $6 million last years, a reduction of 45 percent since 1948, even with an inflationary spiral. And it is keeping a close check on revenues and costs of the remaining trains."

* * * * *

"In the heart of the Allegheny Mountains, on the line of the Chesapeake and Ohio Railway, is one of the great show places of America . . . the famous Greenbrier at White Sulphur Springs, West Virginia. For over a century the Greenbrier and White Sulphur Springs have been revered names in the governmental and social worlds."

* * * * *

"Today, the Greenbrier is still a source of passenger traffic. But the times have created a new leader, the businessman. The Greenbrier is now in large part the coal man's social headquarters, the freight shipper's resort and the businessman's club, a meeting place for important freight and coal shipper organizations. And that is the key to C&O's operation of this luxury hotel."

* * * * *

"Spending $54 million a year to buy some 80,000 different items needed by the modern Chesapeake and Ohio could create endless confusion and delay if systematic inventory control were not maintained."

* * * * *

"Bringing modern management techniques into action, A. W. Hix, Chief Purchasing and Stores Officer, initiated a study of several new inventory systems a number of years just prior to dieselization. By the time C&O received its first diesel, a new simplified inventory control system had been adopted."

* * * * *

"Hauling a big chunk of America's coal down the slopes of the Allegheny Mountains requires a fleet of rugged hoppers. And the C&O has the cars to do this -- 52,608 hopper bottom gondolas. These plus 9,281 flat bottom gondolas, 22,414 box cars and various other types result in a total freight car ownership of 88,481. Thus, the C&O is the nation's third largest owner of freight cars."

* * * * *

"Inevitable, any discussion of freight cars gets around to the subject of roller bearings. To gain operating experience and to check the economic value of them, 1,000 70-ton coal hoppers, built new in 1949-50, and 100 new 50-ton pulpwood cars, built new in 1953, were fitted with roller bearings. All cars were equipped for interchange service. Results to date show no hot boxes, no failures, no maintenance problems."

* * * * *

"Finding that there is more to dieselization than obtaining maximum daily mileages from locomotives, the C&O brought modern methods of business analysis into action."

* * * * *

"Preventive maintenance is basic to the C&O's method of servicing diesels under the balanced load system. The schedules set up provide for periodic inspections of components to detect weaknesses prior to complete failure."

* * * * *

"Rank and file employees are presently given instructions by specially trained supervisors supplemented by a force of diesel mechanical and electrical supervisors."

* * * * *

"The C&O's shops have proved to be another natural area in which it could put to work its advanced management techniques."

* * * * *

"In planning for dieselization, the C&O had a big advantage over many another railroad. By 1950, it was not moving into unchartered territory. The C&O would view the mistakes and the achievements of other railroads. Especially could the Southern Region profit from the experiences of the Northern Region -- the former Pere Marquette -- which had acquired its first diesel, a switcher, in 1939."

* * * * *

"Accordingly, without official authorization beyond consent of mechanical officers to 'see what they could do' but with the cooperation of Work Supervisor Ernest E. Slack, and Shop Superintendent Les Savage, 60 of the men devoted six weeks partially on company time but largely on their own time to the preparation of a big three-dimensional model to sell 'the brass' on how the steam locomotive shop should be converted."

* * * * *

"President Walter J. Tuohy, upon viewing the model, invited the workmen to bring it to New York and present their story before the Board of Directors. Robert R. Young, at that time Chairman of the Board, thanked the men for their work on the project saying, 'On behalf of the board, I wish to express our gratitude for this project which is going to save some three-quarters of a million dollars a year. This confirms the feeling I have had ever since I worked with labor myself, that the men at working level deserve as much of the credit as the officers up the line."

* * * * *

"A substantial portion of the $145 million which the C&O has spent on engineering since 1946 has gone towards building a more stable roadbed for modern railroading. Such work, in turn, is reducing maintenance requirements in places that previously needed constant attention. It also has eliminated costly slow-orders, thereby contributing greatly to on-time train performance and faster terminal-to-terminal schedules."

* * * * *

"The heavy changeover from steam to diesel power has reduced the need for many of the larger roadway improvements that would have been necessary to meet the increased tempo of railroading if steam had been retained. For example, the ability to handle trains without pusher service wiped out, practically overnight, the need for many grade reduction projects."

* * * * *

"Mechanization and motorization has speeded the transportation of men and the movement of materials on C&O's track and roadway forces. Small section gangs and special forces have in many cases been motorized."

* * * * *

"About three years ago the C&O began to intensify the programming and scheduling of its maintenance of way work. Such programming now includes practically all of productive work, such as surfacing and ballasting, tie installation, rail

Chesapeake and Ohio Coal Train.

laying, ballast cleaning, chemical vegetation control and bolt tightening. The approved programs are outlined in a program book, which sets forth the forces and their make-up, location, schedule, materials, and heavy equipment that are assigned for each job."

* * * * *

"Cost-conscious C&O goes in for reclaiming material and machinery in a big way. This helps it get the optimum economic life out of its machinery and equipment investment. It also helps dispose, with the best return to the company, that which has worn its usefulness."

* * * * *

"The largest single location for such work is at the Barboursville, West Virginia, Reclamation Plant. Over 9,000 carloads of material are handled in and out of the huge 40 acre plant in one year. It has 15 buildings and its shops include: work equipment, sheet metal, motor car, rail cropping, frog and switch, blacksmith, boiler, steel bridge, carpenter, paint, and signal. There also is a precast concrete plant and a rail and scrap yard."

* * * * *

"Dieselization, which has progressed rapidly on the C&O in the last few years, has had a pronounced effect upon bridge design and bridge maintenance. It is quite possible that less strengthening and rebuilding will be required to maintain bridges. That's because the diesel imposes lower impact and live load stresses on bridges. Other long-term factors also are studied when the strengthening or the replacement of any bridge is under consideration."

* * * * *

"CTC on the C&O already totals 1,347 track miles. An additional 323 track miles of modified CTC speeds trains on the Russell and Hocking Divisions where full CTC has not heretofore been necessary."

* * * * *

"To tie together its far-flung operations, the C&O is intensively expanding its communications facilities. New carrier telephone and telegraph circuits are increasing the message capacity of its pole lines. Teletype units have further increased message capacity and now form a vital link in an entirely new mechanized system of reporting which promises new speed and economy. Radio, which has proved such a time saver in yard work, is being installed where the Operating Department finds it most useful."

* * * * *

"Unique among railroads is the Research Department of the Chesapeake and Ohio. It is a planned element in C&O's advanced management approach to modern railroading. Reporting directly to President Tuohy, this department is free to investigate any problem on which a scientific approach can shed new light."

☆

THE POST WAR CAST

I got to know four of the men described in this chapter. All of the time that Walter was President, I was in a subordinate position in the Huntington Development Office but I got to talk with him at several meetings and drove him once on a local outing at the Guyan Golf and Country Club. Tuohy was a very charismatic individual with a ready smile and crinkly blue eyes that could signal a warm feeling or bore a hole in you if the conversation turned to serious business. He was slightly built but seemed to have boundless energy. He always wore loafer type shoe and I decided to follow his example. He was an individual who could charm any audience and was a true leader by example and force of personality. He was a faithful Roman Catholic who, along with his good friend, Raymond Salvati, probably had a lot to do with the impressive stone church edifice directly across Route 60 from the Greenbrier Hotel.

Howard Skidmore was a luncheon companion at the officer's dining room in the Cleveland office. Eventually, I got to attend these luncheons when visiting the main office. Although I had no direct dealings with him or his department, it took him awhile to warm up to a young hillbilly mining engineer newly hired by the Coal Department. He was blunt, but very competent in the public relations job when he occasionally came under direct fire from the media. He had a keen eye for people and the best interests of the railway at heart.

Howard was a bright young reporter for the *New York Herald Tribune.* Thomas J. Deegan was a close associate of Robert R. Young who was a leading financier of the post war period with offices in the Chrysler Building in New York City. Howard was brought in to assist Deegan and be his surrogate in the C&O headquarters in Cleveland. Howard was good looking, spoke with a typical New York accent and was impressed by the opportunity to participate in the big time of rail transportation.

Hays Watkins was known throughout the company as an individual who would go places. He was mild-mannered, but very competent to get people to reach consensus in the manner of President Jimmy Carter. It was no great surprise when he was later elected President of the company.

I got to know John Kusik on one occasion when I was requested to drive him and two companions from our office in Huntington to the Charleston airport. Before we left the filling station where my car was parked, he excused himself to go to the restrooms. When he came back he said, "This place is not helpful. First sign says, customers only; second sign says, men only; third sign says, out of order."

John Kusik was to wield a powerful influence on C&O for a long time to come. He was the son of a minister in Fallinn, Estonia. At the tender age of 22 he decided to strike out for the new world by rowing a boat to Finland and eventually made his way to the States. English language was learned while working in a shipyard on the Harlem River in New York. Somehow he graduated with a master's degree from the University of Virginia. He was hired away from the job of controller at a GE subsidiary and rose to become Vice President of Finance and Vice Chairman of the Board. His slight accent was never overcome but his talent and innovations were many. He also brought Hays Watkins into the company.

Chairman Young's campaign for railroad improvements was making headlines but veteran railroaders were very skeptical of his methods. At age 51, he was still a relatively young man and prided himself on hiring young men for big jobs. Thomas J. Deegan, Jr., age 39, was Young's right-hand man in his New York office. He became Vice President in charge of C&O's advertising, public relations and executive committee for the Greenbrier. Young also made him President of the Federation of Railway Progress which he founded to promote rail transportation after losing confidence in the existing national organization, AAR.

The dining rooms of the Greenbrier Suite was the site of the daily officers' luncheon. It was part of an ornate living quarters located between the hotel and office building at the Terminal Tower. It was used to house visiting dignitaries including on one occasion the Prince of Wales. When Walter became president he moved his chair from one end of the long dining table to the middle where he could be more accessible to the other officers. Every luncheon thereafter became a staff meeting where major problems were often discussed into the middle afternoon.

Arthur S. Genet was hired by Young to head up the merchandise traffic department. Only 39, he was a commanding figure hired away from a Manhattan Branch Bank. He was obviously very ambitious and tried to get appointed vice president of operations without success. He finally left to become head of the Greyhound bus line.

Charles R. Hook, 34, was Vice President of Personnel and later Labor Relations. His father was the head of Armco and his wife was Louise Morgan, daughter of the banker who had opposed Robert Young. There seemed to be no ill feeling since she was both beautiful and charming.

Robert W. Purcell, 37, was formerly a lawyer for Young at the Allegheny Corporation and became Vice President of the C&O Law Department. In 1953, he persuaded Young to buy control of Investors Diversified Services. Young put him in charge of it and he became a very rich man.

Ken Browne, 43, was head of C&O's Research Department when he was trying to develop Train X and the Railvan, two of Young's pet projects. He was a brilliant introspective individual and a typical research engineer type. Although the two train types had great promise they never became a reality due to the decline of passenger service.

E. T. Lowmaster was the outstanding exception to the Young movement. He was approaching retirement and had operated the railroad from his office in Richmond, Virginia, for many years. He was a large man with a commanding bearing and was much revered as one of the best ever masters at his job. He tried to get this retirement postponed in 1949 but was not successful in the effort. Another much respected officer, Ross S. Marshall, Senior Vice-President, was slight of build and distinguished looking, belying his years as an operating man on western roads. He also retired at the same time as "Mr. Low."

Source:
Young's Young Men, C&O Magazine, Howard Skidmore, Vol. 2 and 3, February and March, 1992.

☆

FOREWARD

After my discharge as Lieutenant Commander by the Navy in Washington, D.C., in the winter of 1945, I went back to West Virginia University to complete my Masters' Degree in Mining Engineering which had been half completed before my commissioning in 1942. A younger coed that I met at WVU had become my wife in 1944 while on leave, and we sat up house keeping in Morgantown in a small apartment. My bride, Nadine A. Bailey, was the daughter of a store owner in Bridgeport, West Virginia, who had served as Mayor of the city.

Since I needed transporation to complete my field work to finish my thesis on conveyor mining methods at WVU, I enlisted my father's assistance in Pittsburgh to acquire a used car. It turned out to be a 1936 model Ford Sedan in poor, and very noisy running condition that required frequent doses of tractor oil to function.

Waiting for me at the School of Mines was a letter from Dr. Charles E. Lawall, who had been the former director of the school and had been promoted to Interim President of the University when I was a senior in 1940. Ray Salvati, President of Island Creek Coal Company, was a former student of Dr. Lawall and as President of the Board of Governors of the University had a controlling interest in this selection. During the war, Lawall was elevated to President, but terminated in 1944 by Governor Neely because he was a member of the opposite political party. He was quickly hired by the C&O as a replacement for Ira F. Davis, the head of the coal development office in Huntington. Raymond E. Salvati was a close personal friend of Walter Tuohy, Vice President in charge of the Coal Department of the C&O. This connection obviously influenced this turn of events. Ira was past retirement age, but was kept on the job to familiarize Dr. Lawall with the operation of the facility. The letter requested that I come to see him in the coal development office on the seventh floor of the C&O building in Huntington before I acepted a job with anyone else. The office was located at the corner of Fourth Avenue and Eleventh Street, built in 1928 and called the Coal Exchange before the railroad bought it during the depression. The meeting was arranged during May when I made a trip to Huntington with my wife in the Ford jalopy. After the interview, Dr. Lawall asked me to drive him to the train station. I tried as graciously as possible to dissuade him due to the deficiencies of my vehicle but to no avail. It was raining cats and dogs and I was humiliated because Nadine had to hold a towel over his knees to catch the water leaking under the dash board. He had offered me a job in the office with a salary of $375 a month.

After my interview with Dr. Lawall, I called my father who was a mining engineer with the Bureau of Mines in Pittsburgh to get his opinion. His reaction was to suggest that I call back immediately and accept the offer. He knew the C&O was a large prosperous railroad and he was a personal friend of Dr. Lawall. He told me that I would be very fortunate to practice my profession with such a fine organization rather than a coal company, many of which were not too prosperous or stable. Needless to say, I called back and made arrangements to report for work.

Monday, July 1, 1946, was to be my first day on the job, so we arrived at our dilapidated apartment right next to the Railway ROW on Saturday. My brother Lee was in Marshall at the time and at my request obtained one ofthe very few available apartments that was barely habitable. I had written to Dr. Lawall and told him where we were staying and received a phone call from him on Saturday evening saying that I should be ready to leave on a field trip the next day, Sunday, and would be picked up at 1:30 p.m. I had only obtained one suit since becoming a civilian, so I had to report in mostly Navy khakies on such short notice.

The trip proved to be a mission with Dr. Lawall, a senior mining engineer and a field representative, to study the coal outcrops along Peters Creek located near Summersville in order to evaluate the coal reserves in the watershed. The NF&G Railroad which was jointly owned by C&O and NYC was proposing to build a branchline to serve a mine on Peters Creek just south of the Lewis Lands. The C&O could elect to accept a joint venture and put up half of the several million dollar cost. Our party spent five days in the area and took coal samples from many outcrops pointed out to us by a local farmer who had lived in the area all his life. The senior engineer on our team, Guy Hoover, was so meticulous in channeling the coal outcrops, that all of us were completely exhausted by the ordeal. Needless to say, I learned quite a bit about the individuals I would be working with and a "hands-on" study of coal sampling methods.

Out trip was very revealing of the coal reserves in the area.One seam was determined to be incorrectly identified on our map and the reserves were mostly of poor quality and unmineable due to thickness or large shale partings. As a result, management was advised to turn down the offer of half ownership in the branchline, because it would not likely be profitable.

On our return to the office, I was interviewed at length by Mr. Davis. During our conversation he advised me that my job as field engineer, if performed well, was worth $5,000 a year and that is what he had requested for my salary. This was 20% more than my original offer from Dr. Lawall. My title was Field Engineer. No career could be started on a more optomistic basis.

The Coal Development Office was set up by Vice President John Davin in the late thirties to administer the vast coal lands owned by the wholley owned subsidiaries Western Pocahontas Corporation in West Virginia and Chesapeake Minerals, Inc. in Kentucky. In addition, the office handles all applications for branchlines and sidetracks for coal loading purposes as well as a routine mine inspection system to assist the Car Allotment Commission in the Operating Department.

Ira Davis was an engineer with a degree from the University of Virginia, who was most helpful to the Virginian Railway in building their original line into the southern coal fields in spite of C&O opposition. Obviously impressed by his ability, he was later hired away by the C&O to operate their two fuel mines near Whitesville, West Virginia. When I arrived on the scene he was way past retirement age and was in the process of being replaced by Dr. Lawall. He was a very conservative individual and wise in all aspects of engineering in the coal business. He seemed to be much impressed by my willingness to pursue the tough regimen of walking and crawling through coal mines.

Guy Hoover was close to 60 years old, a really briliant engineer educated at Ohio State, and had spent most of his career as chief engineer for Gauley Mountain Coal Company at Ansted, West Virginia. Ed Haessler, field representative, was a hail fellow well met who grew up in New York City and developed his own limousine service. Walter Tuohy has used his services on several occasions on trips to the Big Apple. Walter liked his personality and seeming desire to meet his every wish. Ed was obviously hired to be the boss' personal observer in the Coal Development Office. J. B. Herring was a civil engineer who had worked many years with Darcy Morton in the Car Allotment Section of the C&O operating department and transferred to the coal development office. J. B. graduated from VPI. He was a very intelligent individual who had advanced near the top as a cadet in the Army ROTC Corp, and brought all of his military demeanor with him to the C&O.

The coal loading applications were processed by two field representatives in the office. Ed Heassler covered the Coal River Field, the New River Field centered in Beckley and the NF&G Railway territory north of the New River.

Earl Kiser was a "special representative" in the office who had helped to start the Coal Development Office while a clerk and secretary for John Davin, former vice president of the coal department. John was greatly revered by most of the coal operators on the railroad for his efforts to develop a fair and workable car allotment system; inaugurating the policy of paying a premium price for coal and even-handling of the orders among all the coal companies. Earl was his secretary and understudy in all of this and when John was elected President of the Nickel Plate Railroad, Earl was elevated to a management position as "special representative." He was a superb agent for the Company in having the respect and trust of the operators in the Logan and Kentucly fields where he spent most of his time.

Dick Waybright was a good-hearted individual who had come up through the ranks to be Chief Clerk in the office. He was most helpful to me from the very beginning. He was somewhat of an extrovert and enjoyed throwing his weight around in a very adroit manner. He was plagued by a chronic case of cirrohous on his upper arms and scalp which gave him lots of annoyance.

Three male clerks rounded out the office staff. Jimmy Ward was a little man with a round face and an outgoing irascible nature. He always wore a vest with a pocket watch. As Mr. Davis' secretary, he was very impressed with his own importance. He also maintained the office files in such a manner that he was the only one who could retrieve information with any alacrity. He was in his fifties and tried to denigrate a young inexperienced engineer, such as myself. Russell Blue was another small man in his late forties, a good secretary soft spoken and accommodating in manner. Warren Carper manned the switchboard for the office and performed typing duties. He was in his fifties and his countenance reflected poor health, but he performed competently on the job.

The head of the coal development office reported to Walter Tuohy in Cleveland, Ohio. His office as well as the other segments of the C&O headquarters were located in the Terminal Tower. It is a magnificent 52 story building which contained the railroad passenger station and rapid transit terminal in the basement and also included the Cleveland Hotel and Higbee Department Store. Located in the heart of the city, its ornate spire is easily recognized from the air and is the dominating edifice on the city square.

☆

1946 - THE POST WAR ERA BEGINS

When I reported for work for the C&O in 1946, all I knew about it was that it was considered to be a prosperous railroad. I had grown up in a small mining community called Erie on the short line of the B&O Railroad in Harrison County, West Virginia. The B&O was known to be having financial difficulties. Little did I realize what a powerful leader the C&O was in the rail industry. The following excerpt taken from the 69th Annual Report dated March, 1947, gives an impressive account of the transportation leader in action:

"The Chesapeake and Ohio Railway has reached its present size as a result of gradual growth that began more than a century ago and is still continuing. Today, it is a 3,106-mile railroad system that links the rich bituminous coal fields in West Virginia and Kentucky with heavy coal-consuming areas in the Middle West and on the eastern seaboard. It extends from Norfolk-Newport News and Washington across Virginia, West Virginia, Kentucky, Ohio, and Indiana to Toledo, Louisville, Cincinnati, and Chicago. The C&O's heavily traveled main lines between Newport News and Toledo are largely double-tracked and are equipped with modern signaling and train-control devices that make it possible to operate with maximum safety and efficiency."

* * * * *

"While C&O has pioneered many fundamental improvements in passenger service since it inaugurated the first truly air-conditioned long-distance train in 1932, your company derives nearly nine-tenths of its revenue from hauling freight. Although shipments include almost every article of commerce from airplane parts and automobiles to vegetable oils and zinc, in normal years C&O carries approximately twice as many carloads of soft coal as all other kinds of freight combined. Chesapeake and Ohio acquired control of the Pere Marquette Railway Company in 1929. Since then, coal movements from the C&O to the Pere Marquette have trebled, while C&O has gained direct connections with the manufacturing centers of Michigan and the railroad car-ferry routes across Lake Michigan. The popular new streamliners that started service on the Pere Marquette last summer are further evidence of the benefits it is deriving from affiliation with the C&O. Merger of the Pere Marquette into the C&O, proposed by your Board of Directors in 1945 and outlined in that year's report, was recommended on October 29, 1946, by the Bureau of Finance of the approved stockholders of both roads. The merger plan awaits final approval of the Interstate Commerce Commission."

* * * * *

"Consolidating these two closely related railroads will add, to the benefits of common control they already enjoy, economies from unified operation which within five years may amount to $5,000,000 annually."

* * * * *

"Chesapeake and Ohio also owns a controling interest in The New York, Chicago, and St. Louis Railroad Company, commonly known as the Nickel Plate. Both Nickel Plate and Chesapeake and Ohio hold substantial investments in the Wheeling and Lake Erie Railway Company."

* * * * *

"Your company had acquired, by March 6, 1947, 387,700 shares or per cent, of the common stock of The New York Central Railroad Company at an aggregate cost of $7,356,028. All of the shares acquired have been or will be deposited in trust, in exchange for voting trust certificates, with the Chase National Bank of the City of New York, as trustee in an agreement with your company and Alleghany Corporation dated June 14, 1945, which agreement was entered into pursuant to an order of the Interstate Commerce Commission."

* * * * *

"In the opinion of your management it would be in the public's interest if the railroads in Eastern territory should eventually comprise three groups -- built around the Chesapeake and Ohio-New York Central, the Pennsylvania, and the Baltimore and Ohio Railroads -- each group having two lines between the New York area and Chicago, and each having access to the rich bituminous coal fields of the Pocahontas region, the Atlantic Coast and Great Lakes ports, heavy industrial producing areas, large centers of population, and gateways to the West and South."

* * * * *

"Since 1899 your company has paid its owners dividends every year with the one exception of 1915. During these 47 years, a total of $514, 597,334 was distributed to shareholders. Your company is one of the few major railroads whose dividends were not interrupted during the depression that preceded World War II."

* * * * *

"World War II exerted so severe and so prolonged a strain on the nation's economy that its effects could hardly be expected to vanish with the signing of a surrender document. In 1946 these effects were revealed in terms of shortages, labor unrest, and higher costs."

* * * * *

"Despite these conditions, your company during 1946 carried 95,625,375 tons of freight and received $193,407,931 operating revenues-more than in any previous peacetime year. Net income was $27,726,780 or $3.62 per share of capital stock."

* * * * *

"Years of complaints had failed to remove this barrier (changing trains in Chicago) that so far as the railroad traveler was concerned, separated the Eastern United States from the West. Deciding to take its case to the people, the Chesapeake and Ohio Railway, early last year, ran a series of full-page advertisements calling attention to the problem. Your railroad offered to adjust its own passenger train schedules to operate through cars between east and west, and called on other railroads to do likewise."

* * * * *

"The C&O has acted as well as spoken for improved passenger service. Last August your company was instrumental in putting two all-coach streamliner trains in service on our affiliate, the Pere Marquette Railway, between Detroit, Lansing, and Grand Rapids, Michigan. From August10, to the close of the year, these modern 'Pere Marquettes' carried 159,504 passengers, compared with only 90,708 fares transported between these points in the corresponding 1945 period on old style trains. With a view to improving passenger service on its railroad, your company is investing $33,170,984 in modern passenger equipment, which will unclude the steam turbine-powered new 'Chessies' for the Washington-Cincinnati run, and new coaches, sleepers, and diners to re-equip the company's main-line through passenger trains. The 'Chessies,' which, it is hoped, will be in service within the next few months, will offer added deluxe conveniences at no extra cost. In addition to soft, adjustable seats, with individual radio attachments, wide windows, roomy lounges, there will be movies and a 'play car' to keep children occupied and amused on long journeys."

* * * * *

"While passenger operations are more in the public eye, freight operations account for the larger portion of your company's business-89 per cent of C&O revenues in 1946 coming from that source. We are continually striving to improve this service, and, at the same time, to keep its cost down. Your railroad recently placed an order for 1,000 seventy-ton hopper cars to be equipped with roller bearings, the first large-scale purchase of freight cars of this advanced type."

* * * * *

"The C&O is a growing railroad, pushing new lines into newer coal fields which will become sources of additional tonnage. Branches, spurs, sidings, and other tracks, adding 73 miles to your railroad's tracks, were completed in 1946. A 14.6-mile extension of the Sandy Valley and Elkhorn subdivision to reach new mines in Wise County, Virginia, was started during the year, and a 22-mile extension from Wayland, Kentucky, into Letcher County, Kentucky, to serve another coal-mining operation, was authorized."

* * * * *

"Important steps were taken by the Chesapeake and Ohio during the past year to maintain and develop the high standard of our personnel policy. As of June 1, 1946, an Employees' Supplementary Retirement Plan worked out by a special Pension Commitee of the Board of Directors, was put into effect. Your company had been a member of the Association of American Railroads since its formation in 1934. At that time the Association took over general supervision of several then existing institutions such as the Car Service Division, Bureau of Explosives, Bureau of Railway Economics, and the Railway Engineering and Mechanical Associations. These functions have been carried on with satisfaction to the railroads and to the public. Some of the activities of the Association concerning legislation, public relations, and traffic relations of the railroads have provoked, especially in recent years, criticism and complaints which threatened harmful litigation. The Association has been accused of maintaining discriminatory freight rates. For these reasons your company resigned from the Association October 15, 1946, but arranged to continue by separate agreements its participation in the operating functions as conducted by the Car Service Division and the operational activities mentioned above."

"Subsequent to its resignation from the Association, your company has actively supported the formation of a new organization to represent the public, railroad security holders, railroad labor, railroad management, and financial institutions, as well as individuals interested in the progressive development of the American railroads. Named the Federation for Railway Progress, it will campaign actively for new equipment and improved services,for a balanced wage and rate structure, for progressive management, and for return of free enterprise to railroads. Mr. Robert R. Young, Chairman of your Board of Directors, has become Chairman of the new Federation. In response to public demand, your company has reacquired and is modernizing the popular Greenbrier Hotel at White Sulphur Springs, West Virginia, which it operated for many years before the war. A good will measure in part, operation of this establishment encourages people to travel by giving them a worthwhile place to go."

* * * * *

From this annual report it is obvious that I had joined a really progressive company with a forward looking program. One other aspect became obvious to me in the first year on the job. This was a general feeling among the employees that we were involved in a team effort that was going to be successful. Everybody in the company was going to prosper in the mutual effort. Thus a genuine feeling of mutual respect gave the Company employees a sense of loyalty and responsibility that could be compared to "family."

One aspect of the coal business gave the C&O a terrific advantage in dealing with its coal mining business. The railway was, in most cases, the best customer of the coal mines on the system and purchased a sizable portion of their production to fuel their locomotives and heat their buildings. In order to favor their customers, the railroad usually paid a premium over the market price for all its purchases. It is not hard to imagine the goodwill engendered by this practice and the resulting good feeling for the railway and its representatives.

My job as field engineer turned out to be a routine program of inspecting all the mines on the system to prepare a written report which was forwarded to the allotment commission. The information was used to determine the actual production capability of each mine on the railroad every month. The commission issued a "rating" which was the number of empty cars that a mine was allowed to receive for their production in one day. Initially all the cars were 50 ton capacity and a 50 car rating would enable a mine to load 2,500 tons in one 24 hour period.

In addition to this regimen, I was assigned to inspect the mines operating on lands leased from the railroad subsidiaries. The purpose of these inspections, was to ascertain if prudent mining methods were being followed and a reasonable amount of coal was being recovered without jeopardizing the future minability of the remaining reserves. Each lessee mine was inspected twice a year.

My schedule was usually left to my own discretion and I would call a mine on the telephone on Friday afternoon or Monday morning and arrange my schedule for the coming week. It was only possible to visit one mine a day, but for the lessee mines, the procedure was more intensive, which sometimes required several days for one mine. Ordinarily I would spend three days and nights a week arriving back in the office Friday to work on writing up the reports. For the first couple of years, we also worked on Saturday morning and were free to go home after lunch.

BOARD OF DIRECTORS

James G. Blaine . . . . New York, New York
Robert J. Bowman . . . . Cleveland, Ohio
Robert J. Bulkley . . . . Cleveland, Ohio
Cyrus S. Eaton . . . . Cleveland, Ohio
Howell B. Erminger, Jr. . . . . Chicago, Illinois
Herbert Fitzpatrick . . . . Huntington, West Virginia
Harvey D. Gibson . . . . New York, New York
Henry J. Guild . . . . Adams, Massachusetts
Allan P. Kirby . . . . Morristown, New Jersey
William H. Lipscomb . . . . Leesburg, Virginia
Robert W. Purcell . . . . Cleveland, Ohio
Harry C. Thompson . . . . New York, New York
Andrew Van Pelt . . . . Philadelphia, Pennsylvania
Robert R. Young . . . . Newport, Rhode Island

OFFICERS

Robert R. Young, *Chairman of the Board* . . . . Cleveland, Ohio
Robert J. Bowman, *President* . . . . Cleveland, Ohio
A. T. Lowmaster, *Executive Vice-President* . . . . Richmond, Virginia
R. S. Marshall, *Senior Vice-President* . . . . Cleveland, Ohio
Robert W. Purcell, *Vice-president-Law* . . . . Cleveland, Ohio
Walter J. Tuohy, *Vice-President-Coal Traffic & Development*
. . . . Cleveland. Ohio
Arthur S. Genet, *Vice-President-Traffic* . . . . Clevalnd, Ohio
William H. Wenneman, *Vice-President-Finance & Corporate Relations*
. . . . Cleveland, Ohio
Edward M. Thomas, *Vice-President* . . . . Cleveland, Ohio
H. P. Henshaw, *Vice-President* . . . . Huntington, West Virginia
John B. Parrish, *Vice-President* . . . . Richmond, Virginia
Henry F. Lohmeyer, *Secretary & Treasurer* . . . . Cleveland, Ohio
O. M. Hepler, *Comptroller* . . . . Cleveland, Ohio

The first real friend I made on the railroad was Arden Trickett, a staff officer in the Real Estate Department. He was an extrovert and a hypochondriac. Even so, he was a very agreeable person raised in Preston County who had graduated in Law at West Virginia University and this gave us a common interest. He had a co-worker in his office by the name of Mark Hankins who was a garrulous and wise land agent raised in Grennbrier County and possessed of much lore of the hills.

Arden complained about the morale in his department. He contended that their boss and the chief clerk didn't ever speak to each other. This was a most unusual situation in a company noted for excellent employee relations and morale.

☆

1947 - MERGER WITH PERE MARQUETTE

One of the highlights of the year was the trial of a new coal burning turbine locomotive developed to pull the new *Chessie* daylight passenger train planned for mid-1948.

Another big event during the year was the reopening of the Greenbrier hotel with a celebration for the coal operators and other invited guests, which almost became a disaster when most of the service personnel at the hotel went out on strike. Middle management people all over the system were called in for temporary duty to save the day. Ed Haessler and Earl Kiser from our coal development office became waiters for the occasion.

An *expediter* fast freight service was inaugurated between Chicago, Cincinnati and Newport News. New lightweight streamlined passenger equipment was bringing back customers on the Pere Marquette with diesel power.

"For transporting--134,342,939 tons of freight, an avergae distance of 29, 156 miles, the Chesapeake and Ohio received $286,418,886 in freight revenue during 1947. Supplemental by passenger and other miscellaneous revenues. Total operating revenues were $312,953,036, an increase of 28 percent over 1946."

"On June 6, 1947, the properties of the Pere Marquette Railway Company were merged with those of the Chesapeake and Ohio Railway Company and became the Pere Marquette District of the C&O."

"Thus did the annual report confirm the two major developments of the year just passed. Robert Bowman was the President at the time and the Company was riding a great wave of prosperity that started during the war. The annual meeting was held in Richmond on April 20, 1948. At that time, there were a total of 41,769 employees on the payroll.

The company handled 93 million tons of coal in 1947--a new record. It also handled nearly 10 billion ton-miles of merchandise freight representing a 19% increase over 1946. The revenue picture follows:

Coal and Coke	$168M	42% increase
Merchandise	$118M	19% increase
Passengers	$12M	25% decrease
Other trans. rev.	$15M	36% increase
Rent from equipment	$6M	13% increase
Dividends	$1M	4% increase
Other income	$1M	40% decrease
total	321M	27% increase

Nineteen million tons were dumped at Newport News and seventeen million at Toledo. In one day, October 30, 1947, 5,911 cars of coal were loaded at mines equal to a train 45 miles long.

Two hundred and sixty-five million dollars were expended for expenses and 78 million were paid in taxes leaving a net income of $35 million. The Company paid a dividend of $300 on each share of common stock and $3.50 on preferred.

Robert W. Purcell was elected Vice-Chairman of the Board on April 22, 1947. M. M. Cronk was promoted from the Pere Marquette to be Vice President and General Manager of the newly merged system. John B. Parrish retired on December 31, 1947 and Edward M. Thomas resigned to accept a position with the Nickel Plate.

In November, 1947, the C&O distributed its Nickel Plate stock to its common stock shareholders. Then it turned around and acquired 400,000 shares of New York Central to become the largest single owner.

During 1947, the Coal Development Office was very busy locating new mines on C&O. As the junior engineer employee, I had very little to do with this phase of the work, but I was really busy learning the routine of regular mine inspection. One of the facets of the duty was staying in primitive hotels in the coal fields. Sylvia Pettry was a widow who ran a very clean and comfortable hotel in Madison, West Virginia. The only drawback was a common toilet on each floor for either men or ladies. I spent much of my evening time there studying the mail order catalogs which offered me an opportunity to catch up on civilian products after four years in the Navy.

Walter Tuohy, the head of our department came to Huntington to attend a retirement party for Ira Davis at the Prichard Hotel. He entertained the group royally by performing a selection of classical and semi-classical numbers on the violin. No one in our office was aware that he had this hobby and agreed that his rendition sounded most professional and was immensely enjoyed by the honored employees.

It was a major milestone year in my life when my son, Brent H. Toothman, was born in the C&O Hospital on October 21. I had very much wanted it to be a boy and was elated. We had found and rented a very confortable two bedroom apartment on Sixth Street near Eighth Avenue, but deliberately failed to advise the owner that a visitor was on the way. I finally decided to go see her and confess before the event and was relieved that it did not elicit any protest. Due to the tight real estate market in Huntington after the war, landlords were very selective in accepting tenants with children.

I soon learned to know most of the personnel in the Cleveland office who worked directly under the head of the coal department. One of the real characters was J. Speed Gray, who handled traffic matters pertaining to export at Newport News where he had long served in the operating department. Speed was an original southerner who had all the characteristics and was a fun-loving extrovert. He was very partial to office cars and bourbon cocktails. His favorite sideline was telling risque jokes which he had

No. 493 basks in all her stainless steel glory at Charlottsville in November of 1948, ready to take a passenger train to Washington or Newport News. Polished roller bearing rods and half covered drivers gave a feeling of power, the slanted nose an impression of speed. These strangely beautiful locomotives began as F-19 Pacifics built at Richmond's Alco Works in 1926. When they emerged from Huntington shops as extraordinarily powerful L-1 Hudsons for use on *The Chessie*, little remained of their F-19 bodies except the boilers.

neatly recorded in a little black book. His protege was Bobby Burroughs, our traffic representative in Newport News.

In 1950, Bill Thurston was brought to Cleveland headquarters as Special Assistant from our Richmond, Virginia, Coal Sales office to take over the duties of Roy Reidoinger, who had been in charge of obtaining the C&O's fuel requirements for a number of years. Roy had been selected to assume the Coal Traffic Sales duties of our many field agencies succeeding George Reinbrecht as General Coal Traffic Manager. Thirteen years later Roy retired as Assistant to the President on President Tuohy's staff.

When I took over the coal development office, George Sandmann was Special Assistant to the Vice President and later became our department head. Through the years I worked directly with him and other members of the staff that developed into warm relationships; people like Harold Lane, Chief Clerk, a true gentlemen; Bill Weber, a hard worker who was never too busy to accept more. Bill later spent some time in our office to learn more about the coal fields and was very helpful to me on many occasions as he moved up the ladder with added responsibilities. Then there was Nick Bondra, Assistant Chief Clerk who liked to tell me silly stories and always carried a big smile. Some among other workers in the same bailiwick at one time or other were Al Carrick, Bob Knotek, Bill Schroeder, Doug Loughead, Tom Hayes, who later was assigned to my office as well as the Milwaukee and Beckley Sales agencies, and Bill Cleary from "across the hall." All of these were devoted employees and easy to work with.

☆

1948 - TUOHY TAKES OVER THE HELM

No mention was made of the fact in the Annual Report that President Bowman suffered a heart attack and although making a satisfactory recovery, he was replaced by Walter J. Tuohy. This was a very astute move on the part of the board and Mr. Young as late progress on the railway would prove. Robert Bowman was retained on the board and Harvey D. Gibson of New York City was replaced by Tuohy.

The Annual Report was the first under the new regime and the President's letter in the front and other information described the year's events as follows:

"The heavy volume of railroad traffic characterized by the war years continued in 1948, but toward the year-end it became manifest that the peak had been passed. The total volume of freight carried by Chesapeake and Ohio in 1948 was six percent less than in 1947, and we think the indications are clear that the trend will continue down in 1949."

"Decline from the very heavy traffic of recent years was inevitable and expected. Stability of earning power and the capacity to keep abreast of rapidly changing conditions in the railroad industry have always been characteristic of Chesapeake and Ohio. By reducing expenses and continuing to improve the efficiency of operations Chesapeake and Ohio will maintain its eminent position in the railroad industry."

* * * * *

"Prominent among the improvements made to date have been the extensions of our lines to serve new coal mines. This part of the program includes 14 separate branch lines which it is estimated will make available for transportation over our lines 850 million tons of coal."

"Other important improvements in this three-year period include new yard and terminal facilities, a modern high-capacity pier for handling Tidewater Coal at Newport News, Virginia, car retarders and major track installation at Toledo, and running tracks, signals and miscellaneous facilities including 386 miles of new Centralized Traffic Control."

"Chesapeake and Ohio's Operating Revenues in 1948 amounted to $334,555,157 compared with $312,953,036 in 1947, an increase of 6.9 percent. Freight Revenue accounted for $307,999,809, or 92 percent of the total. The volume of freight handled declined from 134,342,939 tons in 1947 to 126,102,104 tons in 1948, or 6.1 percent. The greater revenues were derived from the increases in rates established during the year."

"Effective August 21, 1948, the Interstate Commerce Commission authorized a permanent freight rate increase to replace three interim increases under which the railroads had been operating. The interim increases had become effective on October 13, 1947, January 5, 1048, and May 6, 1948."

* * * * *

"Coal traffic is the backbone of C&O's business. The Company in 1948 loaded 12.22 percent of the total bituminous coal production in the United States. During 1948 some 68-1/2 million revenue tons of coal originated on our lines, and more than

18 million revenue tons were received from connecting roads. Coal traffic was 68.9 percent of total tonnage handled. The year's coal and coke traffic produced revenues of $173,093,255 or 56.2 percent of total freight revenues".

"Total revenue tons of coal carried in 1948 amounted to 86,969,306. This was a decrease of 6-1/2 percent compared with 1947, when the greatest coal tonnage in the Company's history was handled. The decrease was due in part to a 40.3 percent decline in export tonnage. For comparison purposes, 64,563,298 tons were carried in 1941, and 52,517,776 tons were carried in 1939."

"During 1948 the average coal car on C&O was loaded with 59 tons. Each ton was hauled for an average distance of 312 miles. The average revenue was $1.98 per ton, $117.45 per carload. The revenue received for hauling one ton of coal for one mile was about two-thirds of a cent."

"From mines strategically located near the geographical center of its lines, C&O hauls approximately 75 percent of its originated coal westbound and 25 percent eastbound."

* * * * *

"At the 1948 year-end, C&O served 352 active bituminous coal mines compared with 322 mines a year earlier."

* * * * *

"The coal producing industry is in the midst of a $500,000,000 plant modernization program."

* * * * *

"Merchandise freight has become increasingly important to C&O in the last few years. In 1948, 38,473,650 tons of merchandise freight were handled by the combined Chesapeake and Ohio-Pere Marquette, a decrease of 5.4 percent under 1947. Merchandise freight revenue of $134,906,554 was a new record high, comparing with $118,153,753 in 1947. This traffic accounted for 43.8 percent of Total Freight Revenues for the year 1948."

"Although an increase in basic passenger fares of 20 percent in coaches and 14.3 percent in sleeping cars became effective July 19, 1948, revenues from passengers for the year declined 3.7 percent under 1947."

* * * * *

"Under the I.C.C. accounting formula, C&O's passenger operations in 1948 showed a loss of $23,200,000 or $3,000,000 greater than the loss in 1947 when the industry as a whole lost close to one-half billion dollars on passenger service."

* * * * *

"Wages are up 112 percent since 1933 but freight rates are up less than half that much."

* * * * *

"In March 1949 the non-operating railroad employees were granted though arbitration a seven percent an hour pay increase, retroactive to October 1, 1948, and will be put on a forty-hour basic work week with the same pay as for forty-eight hours, beginning September 1, 1949."

* * * * *

"The C&O is in good physical condition. Expenditures for maintenance of way and structures increased 5.84 percent and for maintenance of equipment 12.62 percent over 1947. These increases were caused largely by higher wages and higher prices of materials."

The Chessie, the train that was but never was, as Herbert Harwood has called it in a definitive article on its strange career, is seen here on its one and only demonstration run at Harts Run, W. Va., in 1948 with steam-turbo-electric locomotive Number 502 for power. Young planned the train as a super luxurious daylight operation between Washington and Cincinnati, powered by the technologically new turbines which he hoped would keep coal as the primary fuel for the railroads in challenge to the oil-guzzling diesels. But by the time the cars and engines were on hand in 1948, the passenger boom of World War II was in steep decline, and *The Chessie* never run. Its cars were used for a while on the regular C&O trains, then sold to other lines, with only four remaining on the C&O after 1951.—*C&O Ry.*

* * * * *

"Total maintenance expenses in 1948 were $110,445,934, or 33.01 percent of operating revenues, and in the year 1947 were $100,766,125 or 32.2- percent of operating revenues. The increase over 1947 was approximately 9.6 percent."

"In 1948, accruals for all taxes amounted to $40,740,500. This was $10,949,065 more than the net income available for dividends."

* * * * *

"Operating income in 1948 of $39,915,018 (after taxes) was augmented by net rentals for equipment and joint facility rents of $5,533,367 and dividends and other income of $1,890,332 producing total income of $47,338,716."

* * * * *

"The Company has a net investment of $42,182,114 in the stocks, bonds and notes of and advances to other companies."

* * * * *

"The 400,000 shares of New York Central Railroad stock which the Chesapeake and Ohio holds represents our largest single investment in an outside company. Last May the Interstate Commerce Commission denied the request of the Chesapeake and Ohio to vote these shares of New York Central and refused to permit Robert R. Young and Robert J. Bowman to accept the invitation extended by the New York Central to join its board of directors."

"While the company believes that New York Central has great future potentialities, there will be no proposal of any merger or consolidation of the two companies until such a time as the restoration of earnings and credit position of the New York Central indicate the wisdom of such a course."

* * * * *

"During 1948, $86 million was spent for additions and betterments to roadway and equipment. Among the major items were extensions of branch lines at a cost of $7.4 million; improvements to yards and terminal facilities, including the large classification yards at Russell, Kentucky, $11.7 million; a new coal loading pier at Newport News, Virginia, $4.8 million (estimated total cost $8.4 million), and Centralized Traffic Control and signal installations, $3.9 million."

"Additional equipment received during 1948 included 6,474 freight cars, 75 locomotives, two 1,000 H. P. tugboats, 46 passenger train cars, and 197 second-hand troop sleepers."

* * * * *

"We have had in progress on the C&O for about four years past a comparatively modest program of technical research. We are developing plans for a new type, light-weight passenger train known as *Train X* designed to reduce the cost of passenger service."

"The coal-burning, gas turbine locomotive development, under the sponsorship of Bituminous Coal Research, Inc., in which C&O is a participant, is making satisfactory progress. It is expected that a test locomotive will be in actual operation in 1951."

"Along with the gas turbine project, the Company is analyzing present and future motive power needs. Consideration is being given to all types of power not excluding diesels."

* * * * *

Major Tams was a powerful coal operator in the New River Field when I encountered him in his office at his mine on Winding Gulf Creek. He regarded me as a young "whippersnapper" and did his best to scare me off.

When I came with the railroad, I soon learned that the chief engineer was an important job in the company. He was directly in charge of all construction and maintenance of the tracks and buildings of the railroad. L. T. Nuchols was in this position and although I had no direct contact with him, I soon learned that he was quite a character. He was already past middle age, had long curly gray hair and the manner of a southern gentleman. Fred Burris from the Pere Marquette succeeded Nuchols who was retained to work for the Greenbrier. Rudy Tench who came through the ranks succeeded Burris. He became famous in the company as an expert witness. He had the ability to discuss everything except the question at hand.

The underground mining of coal is a hazardous pursuit. I had several close calls during my career of inspecting mines. At a new mine in Logan County the local manager escorted me underground in a small electric personal car. The main entries were only about one-half mile into the coal seam. When we got to the face where the coal was being removed to advance the main entries, I started to measure the thickness of coal when a big flash of light and smoke suddenly erupted behind us. Two men were tramming a roof drill toward the place we were in when the cable reel developed a short circuit and started to blaze with lots of black smoke. I looked around for my guide and he was gone. I finally managed to get around the fire while the two miners were trying to sever the cable by pounding on it with a sledge hammer. Sparks were flying everywhere while the bare rope against the rail was sizzling and buzzing until the cable was severed. I then started to help the men throw rock dust on the burning cable in the reel and it became obvious that the manager had gone outside and broke the circuit at the breaker. The fire was soon brought under control and extinguished.

In personnel moves, William H. Weeneman was replaced as Vice-President of Finance and Corporate Relations by John E. Kusik, Vice-President Accounting. Thomas J. Deegan, Jr. was listed for the final time as Vice-President of Public Relations with out office in Cleveland. It is interesting to note that C&O was one of the very few major railroads which maintained its headquarters off-line. Evidently most of the officers had become attached to the location while the Van Swerigens were running the show and resisted moving out. The nearest section of C&O tracks to Cleveland, was the Nickel Plate line from Cleveland, 85 miles west to Fostoria where the C&O line from Columbus to Toledo was intercepted.

H. P. Henshaw was an elderly individual listed a Vice-President of Real Estate. His office and staff were located in the same building that we were in Huntington.

By this time in my career I had been thoroughly familiarized with the C&O coal properties by being accompanied by J. B. Herring on frequent field trips to make mine inspections. My mentor preferred to be called by his middle name, Byron, but everybody called him "J. B." Because new automobiles were only available in limited quantities after the war, I finally obtained a black Ford Four Door Sedan and had started making mine inspections on my own to obtain information for the car allotment commission still headed up by D'Arcy Morton. He was very friendly with me from the beginning and I admired him very much. His son, Sydney Morton, was employed as a field service combustion engineer in our departments' Richmond office and advanced to head of the office.

It is significant that no one was appointed to take over Mr. Tuohy's job heading our department. He obviously wanted to continue a close monitoring of the coal department and his stand-in was a former assistant, J. W. "Billy" Bahen. Roy Reidinger, the Coal Traffic Manager, was a rather austere fine-looking man of the same Christian persuasion as Mr. Tuohy. I learned later that he had a big heart and there was a very sensitive human being behind that facade.

Our company owned a coal property near Gilbert, West Virginia, called the Skillet Fork Lands. While inspecting it I soon met a timber operator by the name of J. H. "Buck" Harless who had a small office across the track from our land. He was a very personable individual and chided me for not helping him get a contract on our timber for which I had no authority at the time. He later opened a coal mine on our property and became a very successful coal operator after making a fortune in timber. His friendship has meant a lot to me over the years.

☆

1949 - A TRYING YEAR

The stockholder report started out with bad news as follows:

"In last year's letter to you, a moderate decline in revenues for 1949 was projected."

* * * * *

"However, coal, the life blood of our traffic, suffered severely. Not in a quarter of a century was the stoppage of production in the coal mines equal to that of 1949. That cause alone accounted for the low 1949 earnings and the omission of the dividend last month."

"As this is written, the U.M.W.-Coal Operators contract has just been signed. The new agreement putting the coal mines back to work promises stability of operation of the coal mines for thirteen months assuredly, for two and one-thirds years probably."

"Merchandise freight is of increasing importance. The merger of the Pere Marquette Railway Company was an important step toward planned diversification of business and in this past year demonstrated materially its value to the system."

"Good business for the year is expected, despite the late start. Revenues and net earnings are promising, particularly in view of the permanent economies effected."

* * * * *

"After reaching an all time high in 1948 at $335 million, Chesapeake and Ohio's total Operating Revenues fell 18% to $274 million in 1949. This drop was brought about principally by the irregularity of coal mining, particularly in the second half of the year. The hauling of coal and coke normally accounts for about 70% of he tonnage carried and more than half of the operating revenues of your railroad."

* * * * *

"In the first half of 1949, there was little time lost in mine operations."

* * * * *

"In the entire last six months of the year, coal mines served by C&O worked a total of only 60 days, due to three-day work weeks and no-day work weeks. It was only during the period from November 10 to 30 that mines worked on a regular basis. In October the union mines in C&O territory did not work a single day. C&O loaded only 1,888 cars of coal during that month, as compared with 118,835 loaded in October, 1948. The Company was obligated to furlough 9,000 employees and made all economies possible."

* * * * *

"More than one-third of passenger train mileage was discontinued during the year. However, the mileage involved had been unprofitable and the loss of revenue will be more than matched by the expense reduction. Effective January 11, 1949, the Interstate Commerce Commission authorized an interim freight rate increase of about 6% and effective September 1, 1949, this was superseded by a permanent increase of about 10% above the level prevailing at the end of 1948. An increase of 12% in passenger fares in the Eastern territory became effective December 1, 1949. However, the passenger rates are no higher now than they were in 1923."

* * * * *

Walter Tuohy, Leo Schneider and Tom Deegan, VP Public Relations, C&O at a C&O Recreation Association golf outing at Ridgewood Golf Club, June 14, 1949.

REVENUES IN MILLIONS

	FULL YEAR 1949	1948
Merchandise freight	$122	$135
Coal and coke	130	173
Miscellaneous freight	7	9
Passenger trains (including mail, baggage, express)	15	18
Total Operating Revenues	$274	$335

"These increases brought the general level of freight rates to a point of about 57% above 1940 and passenger fares 50% (first class) and 69% (coach) above that year."

* * * * *

"One of the conspicuous areas where our rates are far below a justifiable level is with respect to mail and other 'head-end' traffic. Head-end traffic embraces mail, baggage and express."

* * * * *

"Railroad mail handling on the existing term has been described by a leading editor as 'the only conspicuous case in which a private business actually subsidizes the government."

"The various interruptions in coal mining adversely affected the C&O not only in its revenue but also in its efficiency since it was impossible to adjust expenses and operations in line with the widely fluctuating traffic demands."

* * * * *

"In view of the indicated temporary nature of reduced operating levels experienced in the last six months of 1949 the maintenance programs were continued, by plan, so that the property would be in shape to handle the increased business that would result when coal mining was again resumed on a normal basis."

* * * * *

"Since the end of the war in 1945 more than 10,000 coal cars have received heavy repairs representing maintenance expense of $10,000,000. This work reached its peak in 1949 with a total of 3,030 coal cars receiving such maintenance. Not since 1944 had there been such a large number reconditioned."

* * * * *

"Cost reduction activities in 1949 shared primary consideration by your management only with efforts to add additional revenue."

* * * * *

"Net Income and Dividends. Net income for 1949 after all deductions was $11,357,897 or 4.1% of gross revenues as compared with $29,791,435 or 8.9% in 1948."

* * * * *

"The sharp drop in earnings that took place in the last half of 1949-total earnings for the year were actually less than those reported for the first five months-prompted your Directors in November to defer the dividend customarily declared for payment January 1, 1950."

* * * * *

"Resumption of normal mining activity should put the Company back into a profit-making status promptly."

* * * * *

This annual meeting was held in the Richmond office on April 18, 1950. Several changes in the officers were made during the past year. A. J. Lowmaster who had been running the trains for several years retired. C. J. Geyer was made Vice-President of Maintenance of Way and Construction in Richmond. C. A. Taylor was made Vice-President and General Manager of the Chesapeake District in Richmond and M. M. Cronk for the same responsibility on the Pere Marquette District. Thus it took three men to handle all the responsibilities directed by Lowmaster.

I gradually began to get to know by telephone and reputation, several key members of the Coal Traffic Department in Cleveland. Lin Quarles and George Sandmann were two clerks formerly in our Huntington office who had been transferred to Cleveland. This was obviously for their past service and knowledge acquired about coal learned in the area where the mines were located. Lin Quarles had been Secretary to Mr. Tuohy and went with him to the President's office. Later he was rewarded for his service by being assigned a middle-management job in the Richmond office.

Historically, officers of the railroad always were assigned male secretaries. One reason was the fact that the officers usually traveled over the system on their own assigned office car. These cars were literally a condensed home inside a railroad passenger car. They consisted of a galley, dining room, several bedrooms and a living room area in the back end, with an observation platform outside. They were always coupled to the end of passenger trains and the public was blocked off from access to them in the passage ways. They were an ideal vehicle in which sales executives could entertain customers.

Harold Lane was the Chief Clerk in our Cleveland office and was an extremely pleasant and capable individual to deal with. I admired him a lot in the way he was able to handle his duties in a diplomatic manner.

In April a new graduate engineer was employed to work in our office. Dr. Lawall went to his alma mater, Lehigh University in Pennsylvania to recruit this new mining engineer by the name of John T. Rudnicki. John's father had been a miner in the anthracite coal field. I shared a double desk with him. He and his wife became good friends. John was an intelligent and loquacious individual who early on showed sign of health problems.

The job of breaking him in with the mine inspection work fell to me and we traveled together for several months. He caught onto the work very quickly.

A typical week on my job would be one in which I inspected our lessee mine, Eunice, just outside of Whitesville, West Virginia, on the headwaters of Big Coal River. I would call the mine office on Friday and tell them that I was coming and would report to the mine office at seven o'clock for the mantrip which is the time the day shift commenced work. I also called the Whitesville Hotel and made a reservation. Mr. Shibley was the retired resident owner and an elderly lady, Miss Barb, was the manager. She had attended a church with my grandmother in northern West Virginia at one time.

On Monday morning, I would check in at the office downtown to check on my mail and leave before noon in my company car. I always carried two suitcases. One for my toilet articles and street clothes and the other for mine clothes. The route of Huntington was Route 10 to West Hamlin where I would turn off on Route 3 to Whitesville, a trip of about 100 miles. After checking into the hotel, I would usually take a walk up and down the one main street and eat dinner at Fritz's Restaurant. His place was very plain and fairly clean. I usually ordered a steak which he had to chop out from the ice cooler in the back room.

The hotel consisted of a lobby and living quarters for the owner and manager on the first floor and three floors of small rooms above. There was only one bathroom on each floor and the fact that I had to get up so early was an advantage in getting the use of the only one on the floor. The rooms were sparsely furnished with small double bed, chair, table, sink and linoleum floor covering. I got up at five o'clock with the aid of and alarm clock, dressed, ate breakfast at Fritz's and arrived at the mine office by six-thirty.

Usually, the Assistant Mine Foreman met me and accompanied me through all seven working sections on two levels. One day was spent in the lower Dorothy Seam mine with three working sections. The coal was about 5-1/2 feet thick and we rode to the section in an empty mine car. I took measurements of the coal thickness at the face and made notes of natural conditions and machinery in use. The coal was mechanically loaded into steel mine cars and hauled outside to the tipple.

The second day was spent in the two smaller mines in the Fire Block seam near the top of the mountain. Reaching the level of the coal was made possible by a small car pulled up the mountain on rails by a hoist and cable. These mines had plenty of head room and one could walk easily underground.

Back at the hotel, I would clean up by washing my face and arms in the sink in my room and the rest of my torso in the tub in the bathroom. Coal dust is very hard to remove from your face and hair and was a tedious job. Vasoline or cold cream was useful to get the coal dust out of your eye lashes and eyebrows.

After dinner most of the guests would sit in the lobby and visit, read, or listen to the radio. It was a hard life, particularly in the winter time if there was bad weather. The third day on these trips would usually be spent in inspecting another mine in the area. Some of the mines were in very thin coal seams, down to only 30 inches in thickness. I carried with me a set of hard rubber knee pads to crawl around in these mines. The hardest mines to navigate were the ones where the coal was 40-60 inches thick. The men who worked in these mines would be able to "duck-walk" bent over on their feet. Since I was 6 feet, 2-1/2 inches tall and not used to the regimen, it was very hard on my legs and back. During the first six years of my career, I estimate that I inspected about three hundred different coal mines. From this experience I learned a lot about the coal deposits and mining methods in southern West Virginia and Eastern Kentucky.

On another occasion while inspecting a mine, I was persuaded to ride a very fast moving belt to get back outside. My guide warned me to be careful of the tight headroom and about the same time my hard hat with the battery light was torn off. I was slightly stunned from hitting some kind of obstruent above the belt line. I recovered in time to recover my light dangling over the side of the belt. Needless to say, I kept my head almost down on the belt until we reached as ample headroom man-station just inside the pitmouth. When my guide yelled at me to jump off the belt, I reacted as quickly as I could but the belt was moving so fast that it threw me against the side of the entry. Fortunately I only received several bruises instead of any broken bones. I always had to be especially careful to avoid unfamiliar hazards.

DIRECTORS

James G. Blaine	New York City
Robert J. Bowman	Cleveland, Ohio
Robert J. Bulkley	Cleveland, Ohio
Cyrus S. Eaton	Cleveland, Ohio
Howell B. Erminger, Jr.	Chicago, Illinois
Herbert Fitzpatrick	Huntington, West Virginia
Henry J. Guild	Adams, Massachusetts
Allan P. Kirby	Morristown, New Jersey
William H. Lipscomb	Leesburg, Virginia
Walter S. McLucas	Detroit, Michigan
Robert W. Purcell	Cleveland, Ohio
Harry C. Thompson	New York, New York
Walter J. Tuohy	Cleveland, Ohio
Andrew Van Pelt	Philadelphia, Pennsylvania
Robert R. Young	Newport, Rhode Island

OFFICERS

Robert R. Young . . . . *Chairman of the Board*
Robert W. Purcell . . . . *Vice-Chairman of the Board*
Walter J. Tuohy . . . . *President*
Robert J. Bowman . . . . *Chairman of the Executive Committee*
Robert W. Purcell . . . . *Vice-president-Law*
Arthur S. Genet . . . . *Vice-President-Traffic*
H. P. Henshaw . . . . *Vice-President*
C. R. Hook, Jr. . . . . *Vice-President-Personnel*
Thomas J. Deegan, Jr. . . . . *Vice-President-Public Relations*
John E. Kusik . . . . *Vice-President-Accounting*
G. O. Beale . . . . *Vice-President-Purchasing*
C. J. Geyer . . . . *Vice-President*
Maintenance of Way and Construction
C. A. Taylor . . . . *Vice-President and General Manager*
(Chesapeake District)
M. M. Cronk . . . . *Vice-President and General Manager*
(Pere Marquette District)
Henry F. Lohmeyer . . . . *Secretary & Treasurer*
O. M. Hepler . . . . *Comptroller*

☆

1950 - BACK ON TRACK

The annual report featured a five year review of fiscal performance and reflected management satisfaction that the problems of the previous year were behind them. The annual meeting was held on April 17 in Richmond:

"For many years the bituminous coal produced at the mines served by the Chesapeake and Ohio Railway has been greater than the tonnage produced on any other railroad. This tonnage at C&O served mines has averaged one-eighth of the country's total production-the greatest volume of bituminous coal carried directly from the mines by any railroad."

"There is every indication that this C&O record will continue-indeed, all signs point toward increased production of bituminous coal for the nation as a whole. In 1950, with uninterrupted mining operations in the last ten months of the year, national production of bituminous coal reached 506 million tons, up 15 percent from the year before."

* * * * *

"While C&O has had fairly wide variations in total loadings from year to year, there has been a remarkable stability in C&O's share of national production, ranging from 10.5 percent to 12.4 percent. It is particularly interesting to observe that in non-war years, the high quality coals on the C&O, which have wide diversity of use, have a larger share of the nation's market."

* * * * *

"Of the freight revenues, $163 million, or 55 percent, came from the transportation of coal and coke. In the year, 63 million tons of bituminous coal were carried from mines on the C&O, compared with 50-1/2 million tons in 1949. This was 12.4 percent of national bituminous production, the highest proportion for C&O since 1934, and preserved for the railroad its traditional place as the nation's largest originating carrier of bituminous coal."

"The remainder of the freight revenues, $133 million, or 45 percent, came from transporting merchandise freight. Merchandise revenues in the year showed an increase of 10 percent over 1949, and, more significant, were nearly two and one-half times the 1936 combined merchandise revenues of C&O and Pere Marquette. Growth of merchandise traffic on the C&O has been the object of a carefully planned program of traffic diversification."

* * * * *

"Your Company in 1950 earned $33,947,092 available for dividends on its Preferred and Common Stocks. After deducting Preferred dividends of $686,022 this was equal to $4.25 a share on the Common stock."

* * * * *

"The good earnings of 1950 were almost $4 million more than the average C&O yearly earnings during the fifteen year period 1936 to 1950."

"Indicative of improved operations in 1950, the Company's operating ratio-the relation of operating expenses to gross revenues-was 69.4 percent, compared with 80.1 percent in 1949. It was the lowest operating ratio since 1944. At December 31, 1950 total cash and Government securities on hand was $63,311,180."

"In 1950 capital expenditures for new equipment totaled $43 million, including 128 diesel locomotive units, 778 freight cars, and 144 passenger cars. In addition,

39 locomotive units and 5,622 freight cars were on order at the year end. Also in the year, capital expenditures to road and structures of over $6 million were made, covering a wide variety of projects, including centralized traffic control (CTC) and locomotive servicing facilities."

* * * * *

"Chesapeake and Ohio's railway operating revenues were $318,676,867 in 1950, greater by $44,718,039, or 16.3 percent, than in 1949. At the same time, 1950 railway operating expenses of $221,010,430 were only $1,631,055, or less than one percent, greater then the year before. These results reflected not only high traffic levels and increased rates in 1950, but also a continued effective operating cost reduction program."

"Transportation savings resulted mainly from dieselization and from elimination of unprofitable secondary and branch line passenger train mileage. In recent years state public service commissions have been more receptive to the removal of local trains which have become increasingly unprofitable with the development of highways. As a consequence, unprofitable passenger train mileage in 1950 was less than 1949 by 1,594,000 miles, or almost 30 percent."

* * * * *

"The C&O in 1950 received total revenues of $296 million from its main business of hauling freight. Only three railroads in the country handled more freight cars than the Chesapeake and Ohio."

* * * * *

"C&O's 1950 results showed the underlying strength of the railroad's modern well-maintained plant, geared for efficient transportation plus the effect of continuous, vigorous attention to costs. The year was the first in which the forty-hour work week, at forty-eight hours pay, for railroad non-operating employees was fully effective."

* * * * *

"Taxes in 1950 of $56 million represented 18 percent of C&O's gross revenues. They were up $26 million, or 87 percent, over 1949. Most of this increase was in added Federal income taxes, which for the year came to $35 million. In all of C&O's history, the 1950 total taxes were exceeded only in the two war years, 1943 and 1944."

* * * * *

"As of March 1, 1951, the C&O had been granted certificates of necessity covering $83 million of the $111 million cost of 10,650 freight cars, 261 diesel locomotive units, two carferries and a CTC installation. Part of the equipment was received in 1950, and the remainder is scheduled for delivery in 1951 and early 1952. Additional applications for certificates so far totaling $37 million have been submitted by the railroad, but not yet acted upon."

* * * * *

"To Improve further its cross-lake freight route between the East and the Northwest of the United States, the C&O in 1950 ordered built two 410-foot Lake Michigan steamer ferries for freight cars, to cost a total of $10 million."

* * * * *

"By using the Canadian Division-Lake route, freight moving between the East and the Northwest and Pacific Coast can by-pass the Chicago railroad switching area. Freight moving to or through Chicago travels over the C&O all-rail route via Detroit."

The S.S. SPARTAN (above) was launched on January 6, 1952, at the Christy yards in Sturgeon Bay, Wis. She was the first of two ships, each measuring 410 feet in length with beams of 60 feet. The S.S. Badger, the second of the $10,000,000 twins was launched September 6 of the same year. The twins were christened in a double ceremony.

LOCH LYNN HEIGHTS, on the south side of the B&O main line at Mountain Lake Park, developed into a resort area with more worldly attractions . . . casinos, billiards and liquid refreshment.

* * * * *

"At a meeting on February 20, 1951, the Board of Directors amended the retirement plan to provide for its being funded to the extent permissible by the regulations of the Bureau of Internal Revenue. This change, which does not increase the benefits payable to employees and which provides for funding over a period of years, was made in order to take advantage of the large income tax savings which the Government allows only under a funded plan."

* * * * *

"During the year Harry Lowmeyer retired as Secretary-Treasurer. He was replaced by J. J. Angalone as Treasurer and J. Howard Keeler as Secretary."

Aside from mine inspections, my own contribution was to head up a program of drilling to determine the quantity and quality of coal reserves on a large property called Lewis Land and Coal Company. It was located on the border between Nicholas and Clay Counties and on the watershead of Twenty-Mile Creek, in the Greenbrier coal field. It could logically be served by a new branch line built onto a C&O local track which was already in place at the mouth of Twenty-Mile Creek. This property was one of the big reasons why our office advised management not to participate in the building of a branch line paralleling Peters Creek by the NF&G Railroad. This railroad was jointly owned by C&O and New York Central who had engaged in a battle to develop the Greenbrier coal field. Finally, the court stepped in and forced a jointly owned railroad to serve the area. Fortunately C&O has been able to obtain most of the coal traffic under this arrangement. The Lewis property could also be served by a branch line off of the new Peters Creek line which was built by the New York Central without our participation.

The C&O was so anxious to protect the potential traffic opportunity that they went to the owners of the Land Company and offered to prospect the property. If results confirmed a viable reserve, they agreed also to lease the property for the purpose of finding a coal operator lessee. I worked about two months on the prospecting along with Gates Engineering Company in Beckley who had a close working relationship with the Development office. Lauren Gates, the owner, was a mining engineer who formerly worked for Mr. Davis at the fuel mines near Whitesville.

A local young man with an ancient model T Ford was engaged to transport us on the primitive dirt roads and he brought in a couple of neighbors to help dig the outcrop openings in the top seam which was the Five Block. These three men were backwoods hillbillies who had very little education, but wise in the ways of their world. One had been in prison and they all had wild and funny tales to tell. They even brought us a pint of home brew to sample their local industry product. After the field work was completed and testing results were available, I prepared original maps to show the two minable seams on the property and illustrate our results.

I had assistance in the prospecting program in the person of John W. F. Rudnicki who had just joined our office in the year before. This was his first experience in learning about rural life in West Virginia. He really enjoyed hearing about mountain life from our three hillbilly prospectors.

We went ahead with the lease and started discussing the possibilities of the property with prospective coal operators. It would require about a 15 mile extension to serve a mine at the preferred site and the C&O was prepared to build the line at its own expense if a responsible operator could be persuaded to develop the mine. This construction cost was estimated at the time to be about $8 million.

During the summer, the railroad held a conference to train younger middle management people about communications among the various departments of the company. These sessions were held all day once a month for three times at the Greenbrier Hotel at White Sulphur Springs, West Virginia. J. B. Herring, John Rudnicki and I were designated by Dr. Lawall to attend these meetings. It was a very pleasant diversion from the regular routine of our office.

The three of us would board train #6 at about 2 o'clock in the afternoon for the five hour trip. We spent most of the trip in the spacious men's room lounge in one of the passenger cars having a cocktail party. The arrival time gave us just enough time to freshen up at the hotel and have dinner. After dinner, we retreated to the old white club where we met with other men attending the conference and concluded the day with after dinner refreshments.

The session started the next day at 8:00 a.m. and was all business until about 5:00 p.m. We then boarded the "George" passenger train at about 7:00 p.m., ate dinner on the train and arrived home around midnight. I had already attended a few other meetings before this and learned to avoid stuffing myself on the delicious food. The first time I went there after joining the company, I overindulged and rode back home with a stomach-ache. The beauty and comfort of the hotel was very impressive to me and fortified my pride in our company. The luncheon buffet at the Casino was unbelievably lavish and required herculean resolve to avoid gluttony.

The poor earnings of the previous year had forced a reshaking of grandiose plans to put a Chessie dayliner train in service. The turbine engine was plaqued with problems and the B&O had already started a dayliner between Washington and Cincinnati with a shorter route and better time schedule than C&O could hope to match. Management also decided that they could no longer ignore the advantages of diesel power. A meeting was called by Mr. Tuohy of the Annual Coal Show in Cleveland to give the coal operators the bad news that all-out dieselization was under way. Members of our Coal Development office who were in attendance including myself, were asked to invite all of the coal company top executives. Ed Haessler overlooked an invitation to Harry Gay and his wife from Logan County. He suffered their abuse from their disappointment for some time thereafter. Needless to say, this was a most difficult assignment for Walter, but he carried it out without puling any punches and with wry humor. The audience reacted with shocked surprise. Some wanted to stand up and protest, but it soon became apparent that the die was cast.

On my early trips to the Greenbrier Hotel, I often noticed the elderly Colonel J. G. Bradley in the Old White Club. He was a descendant of two socially promised families in New England who came to West Virginia as a young man to make his fortune by developing a huge virgin coal property owned by his grandfather at Widen, West Virginia. He built his own railroad to deliver the coal over to the B&O line along Elk River.

A very influential man that I met at a coal meeting in Charleston was Walter Thurmond. He was a coal operator in the Logan Field and later became Executive Secretary of the Southern Coal Producers Association where he negotiated directly with John L. Lewis on labor contracts.

Staff meetings generally were held during first quarter of the year at the Greenbrier - DeVine held only one meeting here-then they were held in Board Room on the 35th floor of the Terminal Tower.

During my early years with the C&O some of the oldest and wisest members of the group were Jimmy Young who headed up the Richmond, Virginia, Coal Traffic Sales office, and his counterpart in C&O's Fuel Service Engineering Department on the same floor, George Ritchie. Jimmy's sales staff at the time consisted of Roger Thweatt, Warren Loury, Tom Fitzpatrick and Henry Layne. Ritchie had Tom Duffy in his office with Minott Brooke, Sid Morton, and Bill Pye working in the field all were considered very talented fuel Engineers.

In those days C&O maintained 14 coal traffic sales agencies and in addition to Richmond, the top guns were Frank Harmon in Chicago, Dick Sydnor in Beckley, Henry Everett in Boston, Jimmy Phillips in Cincinnati, Charlie Rothgeb in Cleveland, Norton Schickler in Columbus, Milt Ross in Detroit, Art Byron in Grand Rapids, Bill Hanes in Milwaukee and Carl Tonneburger in Toledo. Detroit led the solicitation staff with six members.

Most of these men and staff were very competent in their jobs and very friendly to me. Working on the staff, some later became agency heads themselves, like John Dellinger in Beckley, Larry Kellner in Chicago, Dale Luebben in Greensboro, North Carolina, Frank Pattison in Cleveland, Larry Ulonska in Chicago, Jim Blankenship in Chicago, Roger Thweatt in Richmond, Virginia, Brian French in Toldeo. Some served that role in several cities or joined the headquarter staff. Even some of the Fuel Engineers got to combine their technical abilities with their sales talents as Agency Heads, such as Sid Morton, Carl Dennis and Bud Walser. Bill Clearly served on the headquarters staff for many years before heading the Cleveland agencies for a few years before retirement. Bill Cleary was one of the best-liked fellows around. Always busy but never too busy to take on additional work and got it done, working into the night.

Bobby Burroughs came up through the transportation operating ranks to head the Newport News office for export coal and later came to Cleveland to direct the Coal Traffic Sales Division for a few years before returning to Newport News.

Bill Wortmann and Graydon Shaw spent just about all of their many years in our Cincinnati Coal Traffic Sales office. They became very active in the Cincinnati Coal Exchange group and always invited me to their annual golf outings. I greatly appreciated this as it gave me an excellent opportunity to mingle with hundreds of members of the coal traffic as well as our own C&O/Chessie people.

Joe Beatty headed up the C&O/Chessie Coal rate division in Richmond, ably assisted by old timers like Ed Reilley and Clarence Noble. In the mid-fifties, a young ambitious and highly intelligent gentleman with a legal education and experience named Bill Bailes joined the rate office. It was obvious Bill had a bright future with the C&O/Chessie, but after a number of years of impressive performance overtures started to come from our primary coal hauling rival, the Norfolk and Western, and he joined them. In a few years, Bill headed their Coal Traffic and Development department as Vice President.

Bill Baile's Dad, Lyn, was a long time employee of the C&O/Chessie Merchandise Freight Traffic Department. He headed their midwestern territory Chicago office for many years and was endearingly known as "Mr. C&O" in the area.

☆

1951 - FIRST AMONG COAL CARRIERS

The Annual Report opened up with a highlights section captioned with the following statements:

"The C&O among U. S. railroads: First among bituminous coal carriers. Second in number of shareholders. Fourth in railroad cars handled and, although only tenth in total revenues, Fourth in net income earned."

The rest of the report laid out the results of a very good year:

"Yours was one of the handful of railroads whose profits increased in 1951. In fact, your company's net earnings were higher in 1951 than in all but two of the years between 1930-1950. Preferred dividends were earned sixty times over. Net income of $4.80 per common share was 55 cents per share more than the year before."

"The 1951 showing reflected the abnormal overseas export coal movement, as well as efficient handling of high tonnage of both coal and merchandise freight. Record revenues were earned from both classes of traffic."

"Despite this, complacency is unwarranted: Chesapeake and Ohio's income--and the income of the entire industry--has not been permitted to rise to an adequate level."

* * * * *

"In the face of such spiraling wage and supply costs, management's cost reduction efforts and projects to increase service and efficiency could not alone balance the scales. As 1951 closed, the Interstate Commerce Commission had before it all the railroads' request for the remainder of a 15 percent freight rate increase, approximately one-half of which had been granted by the I.C.C. earlier in the year."

"And, even more important in the long view, as the year ended the Congress and the nation had before it the 'Bricker Report,' an urgent call for the overhauling of national transportation policy, particularly as it applies to rate-making, competition and regulation."

* * * * *

"Chesapeake and Ohio Railway operating revenues in 1951 of $368 million set a new record, with both coal and merchandise revenue exceeding the prior record year of 1948."

"A vital element in the 1951 coal traffic volume was the unusual export movement at Tidewater. In the year, 30.9 million gross tons of overseas export coal were dumped by all railroads, 84 percent of it at Hampton Roads ports. C&O handled 43 percent of the total traffic, and half of that portion dumped at Hampton Roads. Of C&O's total cost revenues in 1951, export coal accounted for 26 percent."

* * * * *

"Indeed, C&O has had to cope with a traffic problem resulting from the decline of some formerly vigorous markets for bituminous coal, its basic traffic commodity. Although more bituminous is mined today and this fuel is still the country's major source of heat and energy, instead of providing nearly 72 percent of fuel in competitive markets as in 1920, it now accounts for 45 percent of today's greatly expanded total fuel market."

"That decline has had inescapable repercussions upon rail coal movement. Thus from 1930 to 1950 the percentage increase in non-coal freight revenues for Class I railroads was almost double the percentage increase in bituminous revenue."

* * * * *

"Regarding coal traffic generally and the C&O's future, it is significant that about one-third of total United States bituminous production is from eastern Kentucky and southern West Virginia--the very heart of Chesapeake and Ohio territory."

* * * * *

"More than a ton of bituminous, for which there is no substitute, is required for every ton of steel made. Electric power plants are expected to take nearly 10 million more tons of coal in 1952 than the 102 million they took in 1951."

* * * * *

"Merchandise revenues in 1951, the largest in C&O history, increased $11,407,000, or 9 percent, above 1950, and $9,595,000, or 7 percent, above 1948, the previous record year. There were 1,312,500 cars of merchandise handled in 1951, an increase of 29,992 cars over 1950. It is interesting that C&O handled in the year nearly as many carloads of general merchandise as it did of coal."

"Important in its bearing on the future, C&O Industrial Development Department in 1951 was successful in securing 104 permanent new industries for location on the railroad's lines in 1951."

* * * * *

"In May, 1951, C&O inaugurated a westbound manifest freight train, *The Speedwest,* as a companion to its eastbound manifest, *The Expediter.* The new train, carrying business originating in the Southeast and Virginia to the Midwest and Northwest, has been productive of additional traffic. Complete dieselization of the Pere Marquette District in late 1951 has enabled C&O to improve further through-freight schedules between the Northwest and the Buffalo Gateway. Also during the year construction progress was made on two new Lake Michigan carferries, which will speed through service when they join the C&O fleet in 1952."

* * * * *

"Early in 1951 plans were made for dieselization of all mainline passenger trains on the Chesapeake District, from which an annual saving of $2 million is expected. By the end of the year, twenty-two of the twenty-seven passenger diesel units ordered were in service, and the remainder were delivered in January, 1952. Dieselization of passenger train operations on the Pere Marquette District was completed several years ago."

* * * * *

"In 1947 the C&O was one of the first railroad to embark upon a program of reducing passenger train operations costs by discontinuing unneeded, unpatronized branch-line and secondary mainline trains. Permission of state commissions is necessary to remove such train mileage in most cases. This program had continued without interruption, and in 1951, 43 percent fewer passenger train miles were operated than in 1947. This was done without serious inconvenience to the public."

* * * * *

"Total 1951 dividends on the common stock amounted to $2.25 a share, compared with $1.50 paid in 1950. Following payment of quarterly dividends of 50 cents a share in the first three quarters of 1951, an increased payment of 75 cents a share was made in the final quarter and in the first quarter of 1952."

* * * * *

"Gross ton miles per freight train hour, at 64,300, set a new high record. Freight car utilization, as measured by net ton miles per freight car day, improved and reached the highest point since 1947. Gross ton miles per freight train miles set a new record at 3,900."

* * * * *

"One of the contributing factors to C&O's ability to handle increased loadings in 1951 was the fact the C&O had cars available when needed. On December 31, 1951, the system owned 85,720 freight cars of all types. This included 4,000 new coal cars added in 1951 at a total cost of $22 million, and 1,622 new box cars acquired at a cost of $9 million. The program for 1952 calls for delivery of 3,500 coal cars, 1,750 box cars and 850 other freight cars."

"The accelerated car repair program, begun two years ago, has continued turning out more serviceable cars. Additional capacity was also provided by the more intensive utilization made possible by careful planning, as well as from the substitution of 70-ton hoppers for 50-ton hopper cars."

* * * * *

"At the end of 1951, we owned 895 steam locomotives and 389 diesel units. Deliveries in 1952 will add 180 diesel units to this fleet. The Pere Marquette District became completely dieselized in December, 1951."

* * * * *

"Mechanization of track work, under way for several years, was continued in 1951. In 1947, the maintenance of way program required 15 million man hours. In 1951, a comparable program was performed with 10.8 million man hours, and the work was done better with the help of machines."

* * * * *

"H. P. Henshaw, Vice-President, of Huntington, West Virginia, retired under the company's supplementary retirement plan on December 31, 1951."

* * * * *

"C. A. Taylor, Vice-President and General Manager, Chesapeake District, was appointed Vice-President--operations, Chesapeake District, on November 1, 1951."

* * * * *

"During the year 543 executives of 29 of the country's leading corporations, as well as 245 other individuals, took advantage of the opportunity to undergo complete medical examinations at The Greenbrier Clinic, in the pleasant and relaxing surroundings of the famous White Sulphur Springs resort."

After the prospecting was completed on the Lewis Lands, the owners were anxious to lease the property to a large coal operator to start realizing income. Dr. Lawall knew several men in the Bethlehem Steel management and tried to interest them in opening a mine on the property. They showed a great deal of interest because the Eagle Seam on the property is probably the best quality high volatile coal in the whole world. After a while, they decided instead to purchase several operating mines on the C&O. One of them was the Truax-Traer mines on the head of Cabin Creek and another was the Y&O Coal Company's Van mine on our property located on Pond Fork about eight miles south of Madison, West Virginia.

In order to placate the owners of the Lewis Lands, Dr. Lawall agreed to have our Western Pocohontas Corporation lease the land to give the owners some income while we tried to get a coal operator to take over. Negotiations proceeded for several months and about the same time the lease was ready to be activated, the Lewis owners notified us that Bethlehem had purchased the property. Needless to say, there was some hard feelings on our part for the reason that Bethlehem had not kept us informed of their intentions. However, they were a big customer and we had to cooperate with them. They had a representative from the Bethlehem main office by the name of Jack Smith who was their lead man on coal leases and hard to deal with. Even so, we did our best to stay in their good graces mainly because we did not want them to develop the Lewis Land with a branch line to the New York Central which was much closer to the property. We spent a lot of time over my entire career on this one matter.

The year opened up a whole new venture in my career. Dr. Lawall called me into his office in June to tell me that he had recommended me to take a temporary job in the Solid Fuel Administration in the Interior Department in Washington, D.C. He told me that an old friend of his was in charge of the Solid Fuels Administration office and I would be in the equipment and materials division which was dispensing strategic equipment and materials to the coal industry. He explained that he had recommended D. L. McElory to a similar job in Washington during WWII and from it D. L. had landed a prestigious job as Chief Engineer of Consolidation Coal Company in Pittsburgh. He said he thought that a good career opportunity could open up for me.

I decided to accept the offer and left in early July for the Capitol City while it was sweltering in the heat of summer. C. W. Connor was a retired coal operator who had been manager of the Armco Steel Company Nellis Mine, in Boone County on the C&O. He was in charge of the government organization plan to allocate strategic materials such as copper and other materials to the coal industry during the Korean War.

A coal mining manager from Illinois, Clyde M. Woosley, was in charge of the Equipment and Materials Division and I eventually became the number two man in his division. Clyde operated under the Navy system whereby the number two man is called the Executive Officer and does most of the administrative work. The experience was very educational for me in learning modern office procedures and the devious ways of the great bureaucracy in our nation's Capitol.

In October a new mining engineer graduate, James R. Lilly, was hired by Dr. Lawall on the recommendation of Dean Gerald Spindler of West Virginia University School of Mines. Before I left for Washington, Dr. Lawall had interviewed another W.V.U. graduate, Richard Lucas who accepted his offer, but called back the next week saying he had to decline employment.

He later became Dean of the Mining Engineering Department at Virginia Polytechnic Institute in Blacksburg, Virginia, where he joined the faculty after deciding against working for C&O.

Jim Lilly was not a big man, but he had a good attitude and a fine mind. He was later to become a very good friend as well as an astute business associate. He graduated with the largest class of mining engineers at W.V.U., up to that time, which was packed with veterans who had earned government assistance under the G. I. Bill. Jim was a veteran and older than most graduates which gave him an edge in employment. He had married while in the service and was the father of two children. Jim had a pleasant personality and enjoyed humor.

My friend, Arden Trickett, complained constantly about aches and pains. The rest of us kidded him about his "imagined illnesses." Much to everyone's dismay, he collapsed with a massive brain embolism in the local C&O hospital. His death made a sorrowful impression on me.

☆

1952 - ANOTHER GOOD BUT WORRISOME YEAR

Two national events were to have a powerful effect on the C&O. Dwight Eisenhower was elected President to initiate eight years of steady progress for the free enterprise system. On the other hand, Oyster Creek Station in New Jersey started to produce electricity from atomic energy and throw a dark scare into the coal industry. The annual report revealed concern but insisted on dogged optimism:

"The world is at the threshold of the Atomic Age. Will the atom bring cheaper energy to wipe out all other forms of power? Will coal, oil and natural gas be displaced entirely by nuclear energy?"

* * * * *

"Regarding coal, popular scientific literature has led many to believe that there will be no place for coal in the world of the future. Actually, with electric utilities gearing up to meet a four-fold increase in electric demand by 1975, the industry expects to need more coal than ever before, not less."

* * * * *

"Atomic energy plants themselves, in fact, have created a huge demand for electric power-and with it, more coal. Atomic energy plants now in operation, under construction, or planned will need an additional 20 million tons of coal a year for electric power requirements."

* * * * *

"The railroad maintained its place as the nation's leading coal hauler, and also as coal originator. More than 61 million tons were loaded at mines C&O serves, representing 13.3 percent of the nation's production."

* * * * *

"The electric utilities industry has increased its bituminous coal consumption 40 percent in the last ten years, and a similar increase is expected in the next five years. The industry consumed about 103 million tons of coal in 1952, and it is expected to use about 110 million tons in 1953. There has also been expansion in the chemical and related industries, all of which offer promising markets for coal."

* * * * *

"Overseas coal exports in 1952 declined 21 percent compared with 1951. This was not unexpected, and a further drop in 1953 is sure."

* * * * *

"In seeking the answer to the question, 'What lies ahead for coal?' C&O places confidence in a few basic facts: As long as the population of the country and industrial activity continues to increase, energy requirements will rise, and coal will continue to meet a substantial portion of these requirements."

* * * * *

"Net income of $45 million was outshone in C&O's history only in 1936 and 1941. Equal to $5.66 a common share, the earnings were eighty-six cents higher than in 1951 and $1.41 higher than in 1950."

"The regular quarterly dividend of seventy-five cents a common share was paid, a total of $3. Preferred dividends of $3.50 a share were paid as usual."

"On the good side, the year brought a high level of business. Steady mining operations kept coal traffic flowing. In May a freight increase helped offset rising railroad costs. For the C&O, tax deferments were given on engines, cars, ships and other facilities necessary for the country's defense program."

"On the bad side was the longest strike in the history of the steel industry. Export coal shipments dropped from the record levels of the year before. C&O's own Lake Michigan fleet was tied up by a four-month work stoppage."

* * * * *

"Stereoscopic views of sites are a small part of the new bureau's services. Industrial engineers and experts in economic geography and marketing, with special training in industrial location research, staff the bureau's offices at Huntington, West Virginia, and Detroit."

* * * * *

"A new record in merchandise revenues of $144.6 million was reached in 1952, surpassing the previous record revenues of 1951. Indeed, in the year just past, C&O's merchandise revenues alone were greater than the total freight revenues of three-fourths of the railroads in the United States."

* * * * *

"Bulldozer breaks ground for Sylvania Electric Bulb factory at Winchester, Kentucky."

* * * * *

"Atomic energy commission plant spur track in Southeastern Piketon, Ohio, was completed by C&O in a record 40 working days at cost of $1.1 million."

* * * * *

"Lincoln-Mercury automobile assembly plant at Wayne, Michigan, will bring C&O estimated $120,000 traffic yearly."

"Dow Chemical plant at Sarnia, Ontario, is typical of swift expansion in chemicals, steel and other manufacturing along C&O."

* * * * *

"The Spartan and Badger are fast, graceful twin vessels, described inadequately in Great Lakes marine technology as carferries. Both were launched in 1952, and when the Spartan went into service later in the year it carried the title of world's largest and fastest freight car and automobile carrying ferry, an honor which it will share when sister ship Badger enters service in Spring of 1953."

* * * * *

"In January, 1953, C&O and its wholly owned subsidiary Western Pocahontas Corporation purchased an additional 244,741 shares of New York Central common stock, at a total cost of $6,107,000. Until then C&O had held 400,000 shares of NYC common, which cost $7,593,000 but had a market value on January 21, 1953, of $9,550,000. NYC declared a dividend of 50 cents a share in December, and the $200,000 was taken into C&O's 1952 income."

"In commenting to the press on the purchase of the additional shares of NYC stock, enough to bring C&O holdings to ten percent of the total, President Tuohy said: 'This purchase of 244,741 shares of New York Central common stock evidences our unbound faith in the future of the railroad industry and reflects a further step in our company's planned program of diversification."

* * * * *

"The Federation for Railway Progress, of which C&O is a charter member, has a successful 1952, featured by a Fifth Anniversary Dinner at New York's Waldorf-Astoria Hotel, which more than 1,000 persons attended."

* * * * *

" Field Research is constant activity of C&O coal department. J. W. Bahen, assistant to president in charge of coal trafficking and development, leads field party of Dr. C. E. Lawall and Mining Engineers John Rudnicki and Guy E. Hoover of the Chesapeake and Ohio staff."

* * * * *

"Walter S. McLucas, a member of Chesapeake and Ohio's Board of Directors, died February 5, 1953, at the age of seventy-seven. Mr. McLucas was Chairman of the Board of the National Bank of Detroit."

"When H. P. Henshaw retired the year before, he left more than a legacy. He had persuaded Tuohy to give his son a job in our office.

After a year on the job at the Interior Department of the Federal Government, I had learned that some of my contemporaries there were mainly political appointees, but there was enough other talent to get the job done. The career female secretaries tried to wield an influence far beyond their job description. From an apartment in Arlington, Virginia, I became a daily commuter reading the Washington Post on the bus to and fro. Most of all it was an educational and maturing experience. The most troublesome aspect was the frequent irate phone calls from Representatives and Senators trying to get special consideration for their constituents.

My duties consisted of analyzing requests from coal operators for machinery and strategic materials like copper, all of which were rationed by the government. All of this required written correspondence and it soon became obvious that I would have to dictate my letters to a young, inexperienced female secretary. I struggled mightily for a while, but soon developed a whole series of back letters that I could use as a model by substituting new names and minor details. I never did feel confident of dictating and reverted mainly to long hand when I got back to the railroad.

Dick Barnett was one of the best clerks we ever had in Coal Development. He was hired out of another department while I was away on leave. His best attribute was a quick mind and a willing attitude to do whatever was necessary to further the work of our office. He was determined to get ahead.

It was reported in the local press that Alleghany Corporation, during the year, sold half of its holdings of C&O stock to Cyrus Eaton, a member of the Board of Directors.

1953 - EATON BUYS OUT YOUNG'S INTEREST

The 1953 Annual Report called it a "year of improvement" and stated the facts as follows:

"Nineteen fifty-three marked the fourth successive year of improvement in earnings."

"Net income was just short of equaling the all-time record of twelve years ago."

"Net income increased $3.1 million, although revenues declined $11.6 million from 1952."

* * * * *

"Merchandise freight revenues continued to rise, setting a new record. Now, nearly half of what we take in for freight comes from merchandise, the other half from coal, the results of a concentrated program to diversify our business."

"Coal traffic was good. Aside from the expected drop-off in coal shipments to foreign countries, 1953's coal traffic was only slightly below that of 1952. And C&O continued its leadership as the nation's largest coal carrier--a record it will strive to maintain."

* * * * *

The 1953 report also recounted the departure of the Chairman of the Board, Robert Young, on February 25, 1954, when the C&O sold its 800,000 shares of New York Central stock to Clint W. Murchason of Dallas, Texas, and Sid W. Richardson of Fort Worth, Texas. They became business partners with Robert Young who had persuaded them that a great opportunity to make money was opening up with the future of New York Central. The price of $25 a share provided C&O with a profit of $2.4 million.

The new Chairman was welcomed in the following manner:

"As a Director of the Chesapeake and Ohio Railway since April, 1943, Cyrus S. Eaton has taken a leading part in the guidance and support of management policies."

"By his purchase of Allegheny Corporation's remaining holdings of 104,854 shares of Chesapeake and Ohio common stock, Mr. Eaton further demonstrated his confidence in the future of a property he has come to know intimately over the last decade. Already the largest C&O individual shareholder, Mr. Eaton's additional purchase brought his total holdings to 205,854 shares. He also has a beneficial interest in a company owning 45,000 additional shares of Chesapeake and Ohio common stock."

"To the C&O Chairmanship, Mr. Eaton brings over fifty years experience in an important variety of industries. In his business career, which started with John D. Rockefeller, Sr., he has taken a leading role in the organization and development of great steel, ore and coal mining and manufacturing properties."

"In addition to his C&O position, Mr. Eaton is also Chairman of the Board of Steep Rock Iron Mines Limited; Chairman and President of Portsmouth Steel Corporation and Chairman of West Kentucky Coal Company. He is also Director of Sherwin-Williams Company and Cleveland-Cliffs Iron Company. He is a Trustee of

Walter J. Tuohy
Cleveland
President
Executive and
Finance Committees

Cyrus S. Eaton
Cleveland
Chairman of the Board
Executive and
Finance Committees

the University of Chicago, Denison University and Case Institute of Technology."

"Total operating revenues of $344 million in 9153, although down 3 percent from 1952's total of $355.7 million, were the third highest in C&O history."

"Net income of $48.1 million was the best in twelve years, only slightly less than the all-time record."

"Earning a share of common stock were $6.04 compared with $5.66 in 1952. Regular dividends totaling $3.00 on the common stock and $3.50 on the preferred stock were paid."

"Merchandise revenues of $155.1 million established a new all-time record, surpassing the previous high of $144.6 million in 1952 by $10.5 million. High production of steel, chemicals and automobiles helped set the pace."

* * * * *

"Dramatic evidence of widespread industrial development on the C&O was the location in 1953 of a new major plant at each end of the system. These were the Hooker Electrochemical Corporation Plant, at Montague, Michigan, and, one thousand miles away, the Pan-American Refining Company Plant near Newport News, Virginia."

"Vital to the building up of merchandise traffic revenues on Chesapeake and Ohio has been large-scale improvement of the Cross-Lake Route. This route is C&O's 550-mile strategic portion of a fast-freight line that connects East and West by rails that stretch across the northern border of the nation, bypassing crowded rail terminals to the South."

* * * * *

"Key link to the Cross-Lake Route is C&O's Lake Michigan trainferry fleet, which was joined in March, 1953, by the S. S. Badger, $5,000,000 sister ship to the S. S. Spartan, commissioned the year before. Both 410-foot vessels can carry thirty-two railroad cars as well as passengers and automobiles."

* * * * *

" Centralized traffic control, for which $1,184,000 has been appropriated, is being installed from Saginaw, Michigan to Ludington this year."

"Coal and coke revenues of $163.6 million in 1953, representing 51 percent of total freight revenues, were down 12 percent from the previous year, largely as a result of the expected decrease in export coal traffic."

"National bituminous production in 1953 amounted to 453 million tons, compared with 467 million tons in 1952, a decrease of 3.0 percent. C&O maintained its position as the number one bituminous coal-hauling railroad in the country. More than 56 million tons were loaded at mines served by the C&O, equivalent to 12.4 percent of national production."

"Because of stepped-up European coal production, the total United States overseas export coal dropped from 26.4 million tons in 1952 to 14.7 million tons in 1953. As it was, the C&O still carried almost one out of every two cars of export coal, a total of 6.3 million tons. In 1952, C&O moved 11.6 million tons for overseas export."

* * * * *

"We think that the future of coal is good. Admittedly, the oil and natural gas industries have become formidable competitors in the past decade, but we think that the main force of competitive inroads has been spent. Perhaps not in 1954, but not long thereafter we will see a reversal of the trend."

* * * * *

"When the first experimental atomic energy power plant was put into operation, many thought that atomic power would bring cheaper energy to wipe out all other forms of power. Leading experts on nuclear energy do not agree. Development has barely started and large-scale use of such energy is years away. Atoms will be only another fuel, supplementing the fuels already supplying energy needs."

* * * * *

"Revenues from train and cross-lake trainferry passenger operations in 1953 were $16 million, an increase of 3 percent over 1952."

"C&O continues to look upon passenger service as one of the industry's most valuable assets and at the same time one of its most trying problems. On the C&O, continuing effort will be given to cutting the expense of providing this service."

* * * * *

"New York Central and Chesapeake and Ohio in 1953 entered into a joint arrangement looking forward to the building of passenger carrying cars of lighter weight and lower cost for operation on their prospective railroads."

* * * * *

"C&O maintained its excellence record for keeping passenger trains on schedule. The year's on-time performance was 95 percent for all passenger trains on the system."

"C&O's post-World War II capital improvement of more than $460 million has been of prime importance in our cost reduction and improved earnings position. The money was spent for new rolling stock, locomotives and marine equipment, and for new lines, road additions and improvements."

* * * * *

"At the end of 1953, total long-term debt amounted to $395.9 million, a decrease of $4.2 million during the year. With the peak of our equipment purchases now past, debt reduction will continue at an accelerated pace."

* * * * *

"Total taxes for C&O in 1953 amounted to $41.6 million, a reduction of $8.9 million from the previous year."

* * * * *

"Officers and employees of the Chesapeake and Ohio increased their ownership in the company during 1953 by the purchase of 16,482 shares of common stock."

* * * * *

"From the beginning of 1950, the C&O has reacquired by purchases 32,580 shares of its preferred stock at an average price of $84.85. This represents a total savings of $494,000 from par value compared with a total purchase price of about three million dollars. In addition, the Company saved over $200,000 in dividends which would have been paid on this re-purchased preferred stock."

"At December 31, 1953, the book value of C&O common stock was $57.35 per share compared with $54.20 a year ago, an increase of $3.05 a share."

"Mr. Joseph C. Kauffman was appointed Vice-President and General Counsel in May, 1953, succeeding Mr. R. W. Purcell."

"Mr. A. W. Hix was appointed Chief Purchasing and Stores Officer in October, 1953, succeeding Mr. Guy O. Beale."

* * * * *

I returned to my old job at the C&O in Huntington in July. Dr. Lawall told me that he had a special project for me and I was glad to leave the federal government. My wife was glad to get back to Huntington and we rented an old farm house on East Pea Ridge Road, a few miles east of downtown.

While I was in Washington, Warren Carper, one of the clerks in our Huntington office, died and was replaced by Kenny Lemons who had been working at the locomotive shops. When I returned to the railroad, I was assigned a private office next to Kenny's who manned the telephone switchboard in addition to his clerical duties. He had been a medical corpsman in the Navy during WWII, and I took a liking to him immediately. Under questioning, Kenny told me about learning how to wield the hypodermic needles by practicing on an orange and injecting salt water in another trainee. He was sent with the Marines on the invasion of Iwo Jima and he had some horror tales about his experiences there.

On June 9, 1953, there occurred a tragic event which one would never find recorded in the annual report. This account is taken from the June, 1993, issue of the *C&O Historical Magazine:*

"At 5:20 p.m. on Tuesday, June 9, 1953, C&O's nine year old H-8 Allegheny locomotive No. 1642 was enjoying a new lease on life after four months storage. Hauling eastbound freight (coal) extra No. 1642 east, the big 2-6-6-6 exploded with earth shattering force about 300 feet west of CW Cabin. The train, with 123 loaded and 2 empty cars, was moving at an estimated speed of 20 miles per hour. CW Cabin is about a mile west of Hinton yard."

"Killed in the accident were three of the very well known and well liked five-man train crew: engineer, Wilber H. "Skyjack" Anderson, 62; fireman, Joseph W. 'Joe' Sullivan, 46; and head brakeman Orell E. Richmond, 54. Conductor Lewis A. Kincaid and rear brakeman C. W. Neff, riding in the steel caboose, were not injured."

* * * * *

"The sound of the blast could be heard over most of Hinton. Startled residents in the west end came out to their porches and sidewalks when the force of the explosion shook their homes, particularly on Eighth Avenue. One C&O employee likened the sound to an earthquake."

* * * * *

Dr. Lawall called me into his office on my return from Washington and told me that one of our lessee mines, Eunice, had been shut down due to depletion of its coal reserves. It was the only mine in which the railroad owned all of the equipment and the coal camp as well as the coal reserves and the land. The property was donated to the C&O during the "twenties" in return for building a branch line to it which in turn would serve other properties owned by the donor. The railroad opened two mines in the area, Marfork and Eunice, to supply fuel coal for their engines. My first boss in Huntington, Ira F. Davis, was hired away from the Virginia Railway to manage these two mines.

No. 1642 was built by the Lima Locomotive Works at Lima, Ohio in December, 1944.

Mr. Davis employed an engineer by the name of Lauren Gates and the two men developed a great friendship. When the railroad started to get out of the coal business to engender better customer relations with the coal operators, Mr. Davis was transferred to Huntington to manage the Coal Development Office by Vice President John Davin. Mr. Gates moved to Beckley, West Virginia, and opened his own consulting firm with another individual by the name of Ferguson. When I came on the scene in 1946, Ferguson had died and the firm was called Gates Engineering Company. Mr. Gates' oldest son, Leslie, was my age and started working for his Dad at the same time I went with the railroad. Leslie and I hit it off immediately and have been friends ever since. Mr. Davis was able to give a lot of work to the Gates firm and the close relationship was retained by Dr. Lawall. Leslie had graduated as a civil engineer from VPI in Blacksburg, Virginia, and we always banter back and forth especially during football season when they play WVU.

The Real Estate Department sold all the houses and buildings to two separate local companies who in turn sold the units individually. I spent almost two years selling off the mining equipment. My best customer turned out to be a "junk" dealer in Logan by the name of Joe Fish. He was a very shrewd dealer in used mining equipment. He taught me some good lessons in negotiating and we became good friends in the process.

The last manager of the Eunice mine was a young man named Earl Scholl whose father had also been the manager there before him. He was a very capable mining man and was invaluable help to me in selling all the equipment. We had to work fast because some of the equipment was located in remote locations and subject to being stolen. One lesson I learned from Mr. Fish was to always try to get the other party to name a price first.

One of Dr. Lawall's frequent job practices was to call all five mining engineers into his office to help him compose an important letter that pertained to our work. Since most of these letters were addressed to non-mining engineers, it was rather difficult to describe matters pertaining to coal mining in terms plain enough to be understood. Dr. Lawall would announce the general thought he wanted to express in the letter and ask for suggestions as to how it should be worded. I soon learned that no matter how good any of my own suggestions might be, the two senior engineers, Hoover and Herring, would immediately interject other wording so that no one else could participate. Therefore, these sessions were a waste of my time except for the experience of learning about important decisions that I would otherwise never hear about.

After two hours in one of these sessions, Dr. Lawall would usually state that he wanted the letter to go out that evening and on one occasion he asked Jim Lilly if we had three copies of a particular map he wanted to accompany his letter. When Jim said we didn't have any, we witnessed a rare outburst when he became exasperated and ordered Jim Lilly to always have plenty of copies of all maps on hand, even if he had to order a hundred of each.

A leading coal operator and prominent lawyer in Huntington is Luther Woods. He succeeded his father in operating the Crystal Block Coal Company in Logan County and is also a member of one of the leading law firms in the city.

One of the high points of the year for members of the Coal Development office was the Coal Department's annual sales meeting at the Greenbrier. The purpose of these meetings was to permit all of the officers to meet their counterparts all over the system and to encourage ideas and enthusiasm to solicit more coal traffic for the railroad.

Roy Riedinger, the Coal Traffic Manager, was the ramrod at these meetings and his assistant, Leo Schneider, was delegated to set up arrangements and speakers. Leo was a personable individual who came to work for the C&O in Cleveland shortly after high school as a messenger in the President's office, and had worked up to his present responsibility by hard work and a positive attitude.

The feature at the 1953 meeting was a contest at one of the evening sessions after dinner. The questions were all about facts that applied to the coal traffic aspect of the company. At first, fact sheets were distributed the day before to acquaint us with the background information that could be used in the questions. The contest was held in an open meeting and each individual had to answer verbally so that all could hear the answer. One of the traffic agents received a question about how much coal had been dumped by C&O at Toledo and Newport News the year before. He stammered and stuttered around awhile and finally said that he had studied so many statistics that he couldn't keep them all in mind and sat down in ultimate exasperation. My question was to explain who owned and operated the tugboats at Newport News and at Toledo. I didn't know the situation at the Great Lakes, but I knew the C&O had their own boats at News. It was later explained that the tugs at the Toledo docks were operated by private companies and were contracted to serve the C&O piers as needed.

These meetings were delightful as far as eating and sleeping, but the long business sessions got to be a bore before they were over in the usual three day meetings. Most of the participants took full advantage of the cocktail hour in the Old White Nightclub in the hotel. This made it very difficult to stay alert after a big dinner at the evening sessions. Walter Tuohy and Roy Reidinger were noted as two of the very few officers in the Company who did not imbibe alcoholic drinks. Our own local boss, Dr. Lawall, was also a tee-totaler, but never raised objections to the rest of us.

Cyrus S. Eaton
Chairman of the Board

Walter Tuohy
President

A. S. Genet
Vice President
(Freight Traffic)

M. I. Dunn
Vice President
(Maintenance & Construc.)

J. W. Bahen
Ass't to President
(Coal Traffic & Dev.)

C. A. Taylor
Vice President
(Operations)

J. E. Kusik
Vice President
(Finance)

K. A. Browne
Director, Research
(System)

DIRECTORS

Robert J. Bowman	Detroit, Michigan
Robert J. Bulkley	Cleveland, Ohio
Barnum L. Colton	Washington, D.C.
Cyrus S. Eaton	Cleveland, Ohio
Cyrus S. Eaton, Jr.	Cleveland, Ohio
Roger H. Ferger	Cincinnati, Ohio
Herbert Fitzpatrick	Huntington, West Virginia
M. S. Fotheringham	Steep Rock Lake, Ontario
Fay A. LeFevre	Cleveland, Ohio
William H. Lipscomb	Leesburg, Virginia
Walter J. Tuohy	Cleveland, Ohio

OFFICERS

Cyrus S. Eaton	*Chairman of the Board*
Walter J. Tuohy	*President*
Robert J. Bowman	*Chairman of the Executive Committee*
Arthur S. Genet	*Vice-President-Traffic*
C. A. Taylor	*Vice-President-Operations*
M. M. Cronk	*Vice-President and General Manager*
John E. Kusik	*Vice-President-Finance*
J. C. Kauffman	*Vice-President and General Counsel*
J. W. Bahen	*Assistant to President* Coal Traffic and Development
A. W. Hix	*Chief Purchasing & Stores Officer*
R. F. Browne	*Comptroller*
J. J. Anzalone	*Treasurer*
T. H. Keeler	*Secretary*

☆

1954 - MERCHANDISE REVENUE EXCEEDS COAL

The annual report listed the highlights for the year on the first page revealing the merchandise traffic revenue exceeded coal traffic for the first time. Even though coal traffic was depressed, management hailed other developments as a superficial plus for the railroad:

"For the first time revenues from merchandise traffic exceeded those from coal and coke. This means that today Chesapeake and Ohio's earnings rest upon two bases."

* * * * *

"With its broadened traffic base giving a desirable stability, Chesapeake and Ohio also expects to maintain its leading position in carrying coal. The long-term future of coal, by all standards of reckoning, is bright. A substantial improvement in coal production is expected in 1955."

The highlights are listed as they appeared in the report:

"Earnings a share of common stock were $5.01, third highest in the last thirteen years, compared with 1953's $6.04 and 1952's $5.66. The regular $3.00 annual dividend on common stock and $3.50 on preferred were amply covered."

"Merchandise traffic revenues of $143 million for the first time surpassed coal and coke revenues, contributing 51 percent of freight operating revenues. Coal and coke revenues of $138 million are expected to be increased by 10 percent in 1955."

"Industrial development activities brought 110 new industries to C&O's line in the year. New revenues of $4 million annually are expected from these new businesses, compared with $2.4 million annually from the 114 located on line in 1953."

"Capital improvements cost only $15 million, as against $50 million in 1953 and $85 million in 1952. The large post-war improvement program is behind us."

"Debt was decreased by $24 million. No new debt was incurred. Nineteen locomotive units were purchased for cash."

"Cash and working capital were materially increased. Only at one year-end in C&O's history was working capital at a higher level."

* * * * *

Other special developments in the railroads progress for the year were described:

"Also significant in C&O's 1954 record was a marked improvement in working capital and decrease in debt. Expenditures for capital improvements were only one-fourth the average of the last eight years."

"Costs were under control. As one indication, more net income was produced than in 1951, although total freight revenues were one-fifth less than in that year. Operating and transportation ratios continued to indicate efficient performance."

"These results reflected the effort of an alert organization. Employee teamwork was further strengthened, with almost 2,500 employees now having a shareholder interest in their company's operation."

* * * * *

"The authoritative business and finance magazine *'Forbes,'* in its 'Seventh Annual Report on American Industry, issued in January, 1955, gave C&O top ratings in each field evaluated: Management, Labor Relations, Public and Community Relations and Stockholder Relations. The railroad trade journal *'Modern Railroads'* devoted its November issue to a survey of C&O."

* * * * *

"C&O actively participated in projects to advance and improve the industry's position. Full membership in the Association of American Railroads was resumed in November."

* * * * *

"With good results, C&O people engaged actively in joint railroad community committees for improved public relations."

* * * * *

"Ten years ago, before C&O started its program of active traffic diversification, merchandise traffic constituted only one-third of total freight revenues."

"Merchandise revenues in 1954 of $143 million were $12 million under the all-time record of 1953. However, these 1954 revenues were substantially better then for all the years prior to 1951."

"In the year, 110 new businesses secured sites on C&O's lines. From them the railroad expects to earn $4 million annually."

* * * * *

"But the biggest magnet, aside for the natural resources in C&O's 'Center of Opportunity,' is growing recognition of C&O's service to its patrons."

* * * * *

"What it means most is C&O's readiness to meet competitive situations. Chesapeake and Ohio maintained its position as the nation's foremost originator and carrier of bituminous coal. From mines on C&O, 47.6 million tons of bituminous were originated in 1954, approximately one-eighth of national production. This compared with 56.1 million tons originated in 1953, a reduction of 8.5 million tons, or 15 percent."

"Coal and coke revenues of $138 million, representing 49 percent of total freight revenues, were down 15.7 percent from the previous year."

"National bituminous production, estimated at 392 million tons, was the lowest since 1938. An upturn in production is seen for 1955, with production estimated at 415 to 425 million tons. Large stocks of coal above ground at the end of 1953 contributed to the decline in production. At December 1, 1954, stocks at 71 million tons had been reduced 11.4 million tons, or 14 percent, from the year before."

"Overseas export of bituminous from the United States increased to 15.3 million tons from 1953's 14.7 million tons. C&O hauled 6.4 million tons of this, an increase over 1953, and 42 percent of total export."

* * * * *

"As steel goes, so goes coal. It takes a ton of coal to make a ton steel. A big factor on coal's side for the future, therefore, is the expected continuance of a high level of steel production."

* * * * *

"Electric power generation is the biggest single growth potential in soft coal's future. It is predicted that utility output of electric power may rise to 600 billion kilowatt hours by 1957, compared with 471 billion generated in 1954."

* * * * *

JAMES D. FRANCIS

RAYMOND SALVATI

JAMES H. HARLESS

C. E. "Jim" COMPTON

"In making chemicals and new metals, coal is also a major factor, and it will be even more important in the days ahead."

* * * * *

"An admittedly optimistic forecast on the use of atomic energy in electric power production in 1976 was made in December by General Electric Company, an important factor in atomic power development. It included the statement that introduction of atomic fuel would not upset requirements for coal, gas, and oil. The world, it was said, would continue to use these traditional fuels in ever-increasing quantities."

* * * * *

"The coal companies, meanwhile, have been streamlining their operations with an eye to the future. They have undertaken programs of mechanization. Their reserves are ready for the opportunity that a growing population and diminishing reserves of oil and gas will some day offer them."

* * * * *

Dr. Lawall called John Rudnicki and me into his office in the Spring of the year and told us that the Company wanted us to undergo a Personal Evaluating routine at Western Reserve University in Cleveland. He said it was only being done for young employees who could be considered for promotion and was an opportunity for us. I didn't particularly like the idea of being tested after all the exams I had encountered in my lengthy schooling. John and I both regarded this as a supreme test of character and could not be avoided. I told Dr. Lawall later that after the examination, I would like to go back for an interview and he said it could be arranged.

We traveled to Cleveland on the C&O overnight sleeper from Huntington. The first night in the first class roomettes were usually not very restful. Sometimes the car was too warm and the noise of the swaying car and clacking rails were not conducive to sleep. Consequently, we arrived in Cleveland tired and edgy. Our hotel reservations were for a double room with two beds and we headed off in a taxi for the inquisition. The small rotund bespeckled individual who escorted us to the testing room was thereafter referred to as 'Mr. Peepers,' after the character on TV. He had 10 or 15 other victims sitting in the room taking written tests. He quickly handed us a four page questionnaire and said we had 35 minutes to fill it out. While trying to answer the true and false, multiple choice and other type questions, an alarm bell would interrupt our concentration and Mr. Peepers would tell the rest of us to go back to work while he grabbed the particular individual's papers. This procedure went on for almost two days and was only interrupted by lunch and one interview with an inquisitor and another individual evidently assigned as an observer.

I didn't sleep much during our only night in the hotel because it seemed like I was lying close to Niagara Falls from the noise which emanated from the frequent trips my roommate was making to the bathroom. We both slept quite well on the return train ride back to Huntington, because we were relieved and exhausted.

My return trip to talk to the experts about my tests were very interesting. Essentially they concluded that I was reasonably intelligent and educated and suggested that my best career pursuit would be as an overseas correspondent or in the law. Even though the determination was a far cry from my current employment as a mining engineer, I felt very competent and interested in what I was doing and looked forward to going out to work every day.

There were several interesting personnel developments during the year. Robert Young resigned from the Board and his place was taken by Cyrus S. Eaton, Jr., who joined with his father as the controlling interest in the Company.

Mr. Geyer retired and his post was taken over by M. I. Dunn, Vice-President of Maintenance and Construction. M. M. Cronk was shifted upstairs to a new position of Vice-President Staff in Cleveland and C. A. Taylor took over as Vice-President of all Operations. This man had developed a reputation all over the System as a really sharp operating expert.

Mr. J. W. "Bill" Bahen got promoted to Assistant to the President, Coal Traffic Development. He had been running our Coal Department since Mr. Tuohy assumed the Presidency and was finally getting a little recognition. Although Bill was limited in education, he was a smart little "cookie" who worked like a beaver most of the time.

While in Washington on leave of absence, I had become very much interested in finding out if maybe there was economically mineable uranium ore located in any of our properties or along our right-of-way. This was a persistent rumor in the mining industry that a viable ore body had been located in Jim Thorpe, Pennsylvania, which whetted my appetite. I finally persuaded Dr. Lawall to permit Jim Lilly and me to buy a Jeep and equip it with a scintillation counter and a continuous recording device to make a system-wide survey. Rail wheel attachments permitted us to travel over the system on the track and over area highways.

Jim and I traveled over Route 60 operating the detector all the way to Newport News where we planned to start the rail survey going west. Our search got an early boost when the counter took a big jump on the graph paper as we descended into Rainelle, West Virginia. Needless to say, we were immediately encouraged to quickly obtain rock samples of the outcrop along the highway.

The survey was started in June. Jim and I alternated on a one week schedule to cover the entire system. The word had been passed down to the operating people to cooperate with us, and we tried to keep our true purpose confidential. This resulted in some white lies and skepticism on the part of the Trainmasters who accompanied us in their respective territories. Our main explanation was to the effect that the machine was a new method of testing the rails for defects. Some of them accepted the story with no question, but others were hard to convince.

The survey required almost all of three months. It was in arduous and sometimes uncomfortable journey due to heat and bad weather. We had to be very careful to call ahead for track clearance and were often delayed by other traffic which had priority. The day I arrived in Sarnia, Michigan, to take over the job, the temperature was 105 degrees and a new record. A dip in the lake later cooled us down as my wife and young son had come along for the opportunity to see some of the sights in Michigan.

My most encouraging incident in the whole effort was driving the car back to the motel in Southeast Ohio after a hard day on the rails. The stylus on our recorder jumped clear off the scale just as a light rain storm began. I ran the Jeep up and down the section of highway and continued getting very high readings. Needless to say, I was so sure we had discovered a 'mother lode' that I slept very little that night and dreamed of great prosperity for the Company. When we went back the next day to pin down the source of radiation along the outcrop, we got very little response. It finally dawned on me that we had been measuring the minute radiation particles that were escaping from the stacks of the Piketon Atomic Energy Plant located west of where we made the discovery. Evidently, the rain drops had collected and concentrated the small particles in the atmosphere.

Several rock samples obtained during the project were sent off and analyzed only to reveal that they contained about one-tenth the amount of uranium required to qualify as usable uranium ore.

Fortunately I was much more successful in finding mineable coal reserves in future prospecting ventures.

I met Victor N. Green several times at a small country club at Madison, West Virginia, while he was developing a general contracting business. He later took over the former Armco Tipple at Stickney, West Virginia, and became a successful coal operator.

While I was never shown the figures on the profitability of our coal traffic, there was no doubt in my mind that it was better than most of the merchandise traffic. Three things about coal made a difference. It required no special and varied equipment; coal cars were easier to maintain; and much of it moved to destinations in train load volume. Tidewater and the Great Lakes were probably the most profitable movements for coal.

Ed Morrison was an obvious extrovert in the operating department under M. I. Dunn. He was a tall outspoken individual who smoked cigars. I got to know him while our sons played on the same Little League Baseball Team, and he left the company shortly thereafter.

☆

1955 - LAWALL ELEVATED WHILE EARNINGS CLIMB

The 1955 Annual Report gave the following highlights of the best year in C&O history:

"Best earnings in company history of $7.25 a common share were up 45 percent from $5.01 the year before . . . Effective with the final quarter, the dividend on common shares was increased to 87-1/2 cents from 75 cents, raising the annual rate to $3.50 from $3.00."

"Best operating revenues of $380.3 million were 25 percent greater than the year before . . . Best merchandise traffic revenues of $163.2 million were earned for moving a record of tonnage of general freight . . . Coal and coke revenues of $189.3 million, second highest in company history, were earned for moving coal tonnage which was almost one-third greater than the year before."

"Best cash position of $88 million and best working capital of $49 million were held at the year end . . . At the same time, debt was decreased $16 million, although $37 million was spent on additions and improvements to physical plant."

* * * * *

"This 1955 performance was even more striking when compared with the year just preceding: total operating revenues up 25 percent, merchandise traffic revenues up 14 percent and coke and coal revenues up 37 percent."

"An all-time high net income of $58 million was produced from this record amount of business. This was $10 million more than was earned in the previous record years of 1953 and 1941. Net income in terms of a common share was $7.25, compared with 1954's $5.01."

"In addition to revenues and earnings records, other important all-time highs were reached. Cash positions of $88 million and working capital of $49 million were both at year-end record levels. At the same time, $16 million of debt was paid off, even though $37 million was being spent on new equipment and improvements."

* * * * *

"Not only was business generally good in 1955, but the benefits of C&O's diversification program--the building up of new sources of merchandise traffic in the railroad's territory--were profitably apparent through the year. For a half-century regarded as a 'coal railroad' only, C&O a decade ago embarked on a program to diversify its revenue sources. This has succeeded to the point where today merchandise traffic revenues contribute almost half of total freight revenues."

* * * * *

"116 new businesses located on C&O in 1955, sources of an estimated $7.1 million in annual traffic revenues. This growth brought to 1,200 the businesses which have settled on C&O in a dozen years. And as was the case in previous years, the railroad's customers at new sites in 1955 included many companies from the first ranks of industry: General Motors, Ford, Union Carbide, DuPont, Amoco, Esso and many others."

* * * * *

"Inbound ore traffic, mainly of chrome, maganese and iron ore, at the Newport News piers reached a record of one million tons. This record, however, will be bettered quickly when a new bulk-materials import pier begins operating late in 1955. Additional annual revenues of $4.6 million and up are anticipated from the new pier."

* * * * *

"As a result of 1955's high level of business, reinforced by the new sources of traffic, C&O growth passed another significant milestone. For the first time, C&O merchandise freight trains surpassed the 10 billion mark in net ton miles, a measurement of traffic carried."

* * * * *

"National bituminous coal production of 465 million tons was up 19 percent from 1954's 392 million tons, and was the largest production in three years."

"C&O held its traditional place as the foremost originator, carrier and carrier-for-export, of bituminous. From mines on its lines C&O originated 61.6 million tons, slightly more than its usual one-eighth of national production. This was 13.5 million tons than the year before."

"Export coal movement over the C&O rose sharply to 14.8 million tons, equal to the record tonnage of 1951 and more than double 1954's 6.4 million tons. Total national overseas export was 34 million tons, more than twice 1954's 15.3 million tons."

* * * * *

"A record 70,000 passengers rode C&O trains servicing the Greenbrier at White Sulphur Springs, West Virginia, and the Homestead at Hot Springs, Virginia. Trainferry travel across Lake Michigan also set a new record, with 201,000 passengers and 72,000 automobiles using the shortcut between Michigan and Wisconsin."

"More than 50,000 school children rode C&O trains to Washington and other points of interest in special group tours designed to add realism to a child's education. Chessie Theater Parties to New York from West Virginia, Kentucky, and Ohio cities were also successful in creating additional passenger traffic."

* * * * *

"Passenger traffic revenues, at $15 million, were four percent of total operating revenues."

"The increased volume of freight and passenger traffic was moved with efficiency. This is shown by the rail industry's familiar yardsticks, usually favorable for C&O, in 1955 even better. C&O's operating ratio (the portion of revenues absorbed by expenses) of 68.1 percent was the lowest in eleven years."

* * * * *

"Rolling stock ordered were 13,000 freight cars and 211 locomotive units. Freight car types included 2,500 box cars, 1,500 gondola cars for steel, 500 covered hopper cars, 500 flat cars, 250 pulpwood cars, and 7,750 open-top hopper cars."

"Important additions to plant capacity were made or started for completion next year. At C&O's great port at Newport News, an $8.3 million bulk-materials import pier will begin operation by year's end. The special ores, iron ore and other bulk materials which will flow across the pier will provide return loads for C&O open-top cars which bring coal to Newport News. Starting next fall, some of this coal will be moving over a new $3.2 million coal dumper and loading tower, which will enlarge C&O's ship-loading capacity at Newport News by one-sixth."

* * * * *

"Electronic train signaling--Centralized Traffic Control--was extended to another 210 miles of track, speeding up train movements and increasing safety. This brought total CTC trackage to 1,744 miles."

"C&O started to use the New York Central Railroad Tunnel under the Detroit River between Detroit and Windsor, Ontario. Combined use of the facility will result in economies to both railroads and improve schedules for C&O. C&O carferry service between Detroit and Windsor was discontinued."

* * * * *

"An addition to facilities of interest was the installation late in the year of a Univac electronic digital computer, first large-scale computer to be installed by any railroad."

"As another modern aid to business, C&O in 1955, leased a 'Flying Office Car,' a two-engine passenger plane for use by C&O's executive officers on flying business trips. Painted C&O blue and yellow, and with the railroad's insigne, 'C&O for Progress,' on its nose, the plane will enable the railroad's officers to cover more efficiently over 5,000 miles of railroad operations."

* * * * *

"The year brought changes among the railroad's major officers. J. E. Doyle, formerly Assistant Vice President-Traffic, was appointed Vice President-Merchandise Traffic, succeeding A. S. Genet, who resigned to become President of Greyhound Corporation. Dr. Charles E. Lawall, Assistant Vice President-Engineer of Coal Properties, was named Assistant to the President-Coal Traffic and Development."

"The loss of J. W. Bahen, Vice President-Coal Traffic, was deeply felt by the many in the railroad and coal industries whose respect and attention he held. Fifty years of age at his passing--typically, while on the job, attending a coal meeting--Bill Bahen had been with the Chesapeake and Ohio for thirty-seven years."

* * * * *

"Most striking is the fact that the Chesapeake and Ohio, which traces its history back 170 years, is really a new railroad. Almost all locomotives--900 units--have been acquired since 1945, along with 28,353 freight cars, 204 passenger cars, the trainferries Spartan and Badger, and 1,211 miles of Centralized Traffic Control. In addition 51,200 freight cars have been rebuilt and trainferries Pere Marquette 21 and Pere Marquette 22 fully modernized. C&O's roadbed has been renewed through the installation of 2,225 miles of new rail and 7.5 million new cross ties."

"These and many other additions to an modernizations of railroad property during the post-war decade have cost C&O a total of $520 million. Nearly half of the $1.1 billion of transportation property on the books at the end of 1955 has been added since the end of World War II. As a result C&O is among the most modern and efficient of U. S. railroads."

* * * * *

"Although sixteenth among railroads in mileage, it stands fourth in net income. And among the thousands of corporations in the nation, C&O ranks about fortieth in net income."

* * * * *

"'In earning power and growth of equity, the C&O has few if any peers, according to Forbes, magazine of business and finance, in the January, 1956, issue. [Boasting the East's lowest nine-month operating ratio (66.7 percent) . . . the C&O brought

Typical of coal mines in West Virginia is Island Creek Mine No. 22 at Holden, W.Va., on the Island Creek subdivision of the Logan district, seen here in 1943, populated with C&O hoppers loaded with various grades of coal.
C&O Railway Photo, C&O Historical Society Collection

home a whopping $48-million profit from $315-million, 10-month revenues,' Forbes continued. 'Considering the earnings buildup financed by plowed back profits, its average 50-50 payout policy is just."

"Forbes, in its eighth annual independent evaluation of 'management capabilities' of the leading twenty railroads, gave C&O, which it termed 'Brightest Star in the East,' a scoring of 95. Only two other railroads, one in the South and one in the West, were rated as high."

* * * * *

An important task of the Coal Development Office was to supply an estimate of coal loadings for the next year by the end of October. This estimate was used to compare with one prepared by the Traffic Manager of the Coal Department and from these figures, the Vice President submitted the final estimate to the President. The estimate from our office was much respected because it usually came very close to actual tonnage that ensued during the next year. The one in our office who did most of the work on the estimate was J. B. Herring.

One might suspect that our estimate was derived from discussion with coal companies and major users of coal. While they were always contacted, the major tool in making the estimate was based mostly on an estimate of total business conditions for the new year. Several national publications give a lot of attention to projections of the coal market forecast and J. B. would make a comparison of all of these and finally select the one he preferred. After this was available he contacted several mining publications and other sources to round out background information from which the final figure was developed.

This estimate was very important to the company because it was used by the Finance Department to prepare spending plans for the coming year. Our estimate was usually within 5% of actual loadings.

The elevation of our leader, Dr. Charles Lawall, to the head of the Coal Traffic Department in Cleveland led to several changes in our local Huntington operations. Office car #21 was assigned to the Coal Traffic Department and Dr. Lawall began using it to commute between Huntington and Cleveland. His usual routine developed into being in our office on Monday and Fridays which permitted him to be in Cleveland during the middle of the week. No one was designated to take over the Huntington Development Office in his absence so all of us tried to see him on Monday and plan our work week independently. Guy Hoover was the senior man in our office, but he was an introverted individual and the last one to assume jurisdiction when the boss was away.

Office procedures were somewhat haphazard before and became even more so under the new regime. However, the increased stature of our boss brought some benefits to members of our office. Most of us were accorded the privilege of an annual physical exam at the Greenbrier Clinic. This list included Hoover, Herring, Kiser, Haessler, Rudnicki, Henshaw, and myself.

My first trip to the clinic occurred in March of 1956 when I met Dr. Eugene Morhouse who was about the same age and we got along splendidly from the very beginning. He was a shrewd physician who enjoyed humor and when he started to give me the first proctoscopic exam he said that I should not mind it since I had also been in the Navy.

☆

1956 - FIRST FLASH REPORT ISSUED

President Eisenhower's reelection heralded another record earning year for C&O. The first annual "Flash Report" was made to the Press on the first working day of 1957. It listed the following "Highlights" which were repeated in the Annual Report:

"Net income was $67 million, equal to $8.28 a common share, a 15 percent gain over 1955's record $58 million, or $7.25 a share."

"The quarterly dividend on common shares was increased to $1.00 from 87-1/2 cents, effective with the final quarter, raising the annual rate to $4.00 from $3.50."

"Merchandise traffic revenues were $170 million, compared with 1955's record $163 million."

"Coal and coke traffic revenues were $219 million, compared with 1955's record $163 million."

"Total revenues were $419 million, a $39 million increase over 1955's record $380 million."

"The year ahead looks excellent, with continued growth indicated for both coal and merchandise traffic."

* * * * *

The entire Annual Report was in the form of a letter from the Chairman and the President and contained the following information:

"A year ago 1955 was reported to you as Our Best Year, but all the revenue and earnings records of that year have now been surpassed."

* * * * *

"In 1956, $90 million was spent by C&O in improving and expanding facilities. This was an addition to almost $100 in cash outlays for maintenance. The $90 million was the greatest amount spent in one year to better the railroad."

* * * * *

"Since the end of World War II, over $600 million has been spent on modernization and expansion. Through these expenditures, in 1956 Chesapeake and Ohio became a billion dollar corporation, one of seven railroads of that size in the nation."

* * * * *

"Of vital importance to coal shippers was that not a day's production was lost at a mine on C&O for lack of coal cars."

* * * * *

"For the second year in a row the annual dividend rate was increased."

* * * * *

"More business made more jobs during the year, and average annual wages reached the new high of $5,400."

* * * * *

"For the second consecutive year national bituminous production increased sharply. It topped 500 million tons. This was up 30 million tons from the year before and 108 million tons from two years earlier."

* * * * *

"Originated traffic in 1956 reached a new high percentage, 13.2 percent of all bituminous mined in the country. This was slightly more than C&O's usual one-eighth of national production. Originations were 66 million tons."

* * * * *

"C&O coal carried to tidewater for shipment abroad was an all-time high of 20 million tons, up one-third from the year before."

* * * * *

"As go electricity and steel, so goes coal!"

* * * * *

"Only in the U. S., and in the territory served by the Chesapeake and Ohio, can the nations of the world get the coal they need, particularly metallurgical coal."

* * * * *

"Atomic power in commercial quantities is far off. Atomic power will simply supplement present energy sources instead of replacing them. The opening in 1956 of the world's first large scale atomic power station, which started to feed electricity to a community in Northwest England, vividly demonstrates these facts."

"Britain, because of its shortage of coal and other energy sources, is pushing atomic power development as fast as possible. In spite of this, by 1965 it expects to get only one-eighth of total power from atomic energy. And in that period, Britain's total power needs are expected to double. Obviously, greater amounts of all sources of energy, including coal imports, will be needed."

* * * * *

"At Newport News, an additional coal-dumping tower was rushed to completion in only fifteen months. It will increase the railroad's coal-dumping capacity by a half-million tons monthly in 1957."

"Coal car yards at Newport News and Clifton Forge, Virginia, were enlarged in 1956. Also, twenty-two construction projects, involving tracks serving coal mines, were completed or under way at year end."

"It is of great importance that a coal mine's operations not be halted because of a lack of railroad coal cars to load. C&O acquired 2,327 additional coal hopper cars in 1956 and will add another 7,500 in 1957."

* * * * *

"Continuing the diversification of the railroad's traffic sources, ninety-eight new industries located plants on the C&O during the year. These plants will ship an estimated 27,200 car-loads annually and produce about $3,250,000 in new annual revenues for C&O when in full production."

* * * * *

"DuPont made plans in 1956 to build a plant at Maysville, Kentucky, Dow Chemical will build one at Oriana, Virginia. Kroger Company started construction of grocery warehouses at Detroit and Grand Rapids."

* * * * *

"How good things sometimes come in threes was demonstrated at Montague, Michigan, where Hooker Electrochemical built a plant on top of a thousand-year supply of salt, raw stuff of chemicals. DuPont built next door and makes neoprene with byproducts it gets from Hooker. Union Carbide put up an adjacent plant to supply acetylene, which the DuPont plant also needed. All three gave business to the C&O."

"At the opposite end of the C&O, in Virginia, there was another example of how good railroad business grows. The $35 million refinery of the American Oil Company near Yorktown started operations in December, giving C&O traffic over a 5-1/2 mile spur track the railroad built. During the year, with C&O's assistance in locating the site, the Virginia Electric and Power Company began to build a huge electric generating plant alongside."

* * * * *

"In 1956, 1,241 merchandise freight cars were acquired, and an additional 1,700 were ordered for 1957. Of the new freight cars received, 702 are box cars which bear the insignia DF on their sides, signifying they are equipped with 'damage free' devices which protect the shippers' lading. Another 500 cars so equipped will be received in 1957."

* * * * *

"To handle even greater future tonnages of ores, a new $8 million bulk materials import pier is nearing completion at Newport News. Capacity of the pier will be 8 million tons yearly. Revenues of $4.5 million and up are expected."

"At Russell, Kentucky, a $5.5 million push-button classification yard to sort cars of high-speed manifest freight trains was half completed by the year end."

* * * * *

"In 1956, passenger traffic revenues were $14.8 million, less than 4 percent of total revenues. Travel on the Lake Michigan trainferries produced a record revenue of $1.5 million. Passengers to the Greenbrier, at White Sulphur Springs, West Virginia, and the Homestead, at Hot Springs, Virginia, totaled 71,000, the highest number ever."

"With the addition of 222 diesel locomotives in 1956, the C&O by year end was handling all traffic with diesel power. This was the result of an intensified Dieselization program which since 1951 has cost $133 million in locomotives."

* * * * *

"Total expenditures in 1956 for locomotives, freight cars and other rolling stock were $68 million. Another $22 million was spent on roadway improvements, which included part of the cost of the new ore pier, the new coal pier at Newport News, new yards and the addition of 117 miles of electronic traffic control. At the year end, 1,820 miles of C&O had electronic controls."

* * * * *

"Operating ratio, which is the proportion of revenues absorbed by expenses, was a thirteen-year record low of 67.9 percent despite higher wages and material costs. Gross ton miles per train hour, a measure reflecting both train load and speed, climbed to a record 72,000."

* * * * *

"A two-engine Beechcraft airplane was acquired for use by transportation, mechanical and engineering personnel in inspecting and supervising activities over the 5,100-mile system. The C&O has had a two-engine DC-3 'flying office car' in service since July, 1955."

* * * * *

"Chesapeake and Ohio late in 1956 announced it was acquiring a substantial financial interest in Slick Airways, Inc., nationwide and international airfreight carrier, through an investment in $3,333,000 of Slick ten-year debentures."

* * * * *

"Because ocean shipping has been the tight place in the coal supply line to foreign nations, the coal industry, the miners' union and the railroads joined in an unprecedented partnership in 1956. In June, the Chesapeake and Ohio, the other two Hampton Roads coal railroads, the United Mine Workers and leading coal operators and exporters announced the formation of American Coal Shipping, Inc., a $50 million ship company, to carry export coal."

* * * * *

"American Coal Shipping, Inc., purchased a Liberty Ship, renamed the 'Coal Miner,' which carried its first load of coal abroad before the year end. Thirty additional Liberty Ships from the Government's reserve fleet were applied for and the request was granted."

* * * * *

"Sixteen other general freight and passenger ships were acquired by ACS through purchase of the Bull Steamship Line, one of the oldest freight ship lines."

* * * * *

"Commenting on ACS, *The New York Herald Tribune* said editorially, 'Labor and capital have launched a dramatic cooperative venture in the coal industry."

"At a time when railroad mileage is contracting in this country, the Chesapeake and Ohio added sixty miles to its ownership through purchase of the Washington and Old Dominion Railroad in November. The all-freight operates parallel to the Potomac River in Virginia, near Washington, D.C."

"A further diversification development is the Railvan. Operating tests of two engineering models were concluded, leading to the design of a production unit."

* * * * *

"The Greenbrier Conference of the United States, Canada and Mexico, on March 26, 27 and 28, was the first meeting of the three heads of state ever held in this country. It was also the first time that a Prime Minister of Canada and a President of Mexico had met."

* * * * *

"President Eisenhower wrote in The Greenbrier's guest book that the conference was taking place in the loveliest surroundings of any conference he had ever attended, in war or peace."

* * * * *

"It was also The Greenbrier's busiest year. It was the fifth consecutive year of record patronage, with a total of 60,000 guests. A new outdoor swimming pool was opened in August."

* * * * *

"By the end of 1956, employee-shareowners had subscribed for 61,907 shares, at a total price of over $2.7 million, and had received a half million dollars in dividends."

* * * * *

"M. I. Dunn, a third-generation C&Oer, was named Vice President-Operations, in charge of the combined transportation, mechanical and engineering departments. He had been Vice-President-Construction and Maintenance. Mr. Dunn's father was a C&O attorney and his grandfather was the Chief Engineer and Superintendent of the railroad."

"Mr. Dunn succeeded Senior Vice President C. A. Taylor, who retired after thirty-two years of devoted service and accomplishment in a succession of vital operating and engineering positions."

"Charles E. Lawall was named Vice President-Coal Traffic and Development. Dr. Lawall, who was President of West Virginia University before coming to the C&O, had been Assistant to the President-Coal Traffic and Development."

"A. Witt Hix, who started with the railroad's Purchasing and Stores Department at the age of fifteen, was named Vice President- Purchases and Stores. He had previously been Chief Purchasing and Stores Officer."

* * * * *

"In addition a total of 30,709 shares of common stock were sold to officers and employees under the stock purchase and stock option plans. These shares, plus the 6,870 shares of treasury stock issued in exchange for shares of Washington and Old Dominion Railroad, brought the total of outstanding common to 8,019,463 shares on December 31, 1956."

"Despite the cash requirements in 1956 for plant expansion, debt retirements and the increased dividend, the railroad ended the year with working capital in excess of $45 million and cash and short-term investments of more than $71 million."

"Chesapeake and Ohio was termed an 'industry blue chip' and given top management rating nationally for the fourth consecutive year in the 'Ninth Annual Report on American Industry' by Forbes."

* * * * *

One of my main duties during this period was to act as chauffeur for Dr. Lawall whenever he traveled where the office car would be impractical. His wife, Margie, often accompanied us on these trips and sat in the back seat. She was a most gracious and refined lady but she was almost a head taller than he and may have weighed more. Her weight confirmed her hearty appetite which in turn must have discouraged her husband from eating. They would no more than get in the car to start out when she would ask me where we would eat lunch. Dr. Lawall would almost cringe at this discussion. From long experience traveling in and around our state, I usually knew the best places.

Many of our trips were made to the Northern part of the state and we always traveled up Route Two along the Ohio River. One of the best eating places was the dining room of the Marietta, Ohio hotel, and another was the basement cafeteria in the hotel in Parkersburg. For lunch, Mrs. Lawall would always order a full course meal and I would have one also. Dr. Lawall usually ordered dry cereal or a salad and it was plain to see that he didn't relish watching us eat so heartily.

One day I was driving Dr. Lawall alone and we had finished reciting most of the small talk of the day, when he turned to me and said, "Fred, why don't you take over the Huntington office and run it." I was stunned by the statement and couldn't think of any appropriate reply. Nothing would have pleased me more than to move up to head of the office, but I knew instinctively that to do so without proper credentials would not be fair to my fellow workers and they would undoubtedly resent my arrogance and retaliate in many unsettling ways. I figured that Dr. Lawall wanted to avoid the onus of wearing the "hair shirt" by appointing me properly and having to suffer the resentment of the other engineers in the office, particularly the two who were much older than I. As a result of my inability to make any reply, we rode along for several minutes with no sound except the sound of the engine and the wind rushing past the car. Finally I couldn't stand the silence any longer and I blurted out, "Where do you want to eat lunch?" Nothing more was ever said about this matter.

One of the highlights of the year was the receipt of an invitation to attend the Annual Rodeo at the Greenbrier Hotel on the last weekend in September. This was a weekend party started in 1923 by a coal operator, Holly Stover, who had a mine on the Virginian Railway. He had gradually increased the scope of the affair to include the owners and top management of most of the coal companies in the New River Field as well as railroad executives and other coal related companies.

The meeting started on Thursday with most guests playing golf or tennis on Friday and Saturday. The C&O sponsored an elaborate cocktail party on Saturday evening before the formal banquet where men wore tuxedos and the ladies dressed in long gowns. The N&W Railway provided the cocktails on Friday evening and eventually the L&N Railroad chimed in with a brunch on Sunday to end the festivities.

This event was by far the most prestigious gathering of coal people anywhere in the eastern coal fields. Holly Stover was retired and getting feeble by the time of my first visit and he rode around the hotel in a wheelchair. The C&O realized that the gathering was a terrific customer relations builder and took over the arrangements of the event even before my first visit.

An incident occurred at one of the Rodeo meetings which illustrates the general atmosphere of good fellowship which was reflected by the participants. The men were dressed in tuxedos and the ladies were elegant in their long evening gowns. After the banquet, most of the guests retreated to the Old White Club which was merely the grand extension of the evening festivities.

On this occasion, the place was packed and the buzz of music and conversation was at a peak level. The two small dance floors were so crowded that several couples were dancing off the edge of the hardwood floor. John Straton, who was a good friend of mine and an associate of Leslie Gates, headed for the men's room along one of the narrow aisles. All at once a very attractive lady in a long flowing white gown came toward him down the same aisle and passed out, falling backwards right in front of John. He reached down and grabbed her by the shoulders and pulled her back up on her feet. She opened her eyes and looked up at John and said, "Did you want to dance?" John nodded in agreement and led her onto the dance floor where they twirled until the music stopped and John escorted her back to her table nearby.

The Governor of West Virginia appointed a blue ribbon commission to study and recommend ways the state could improve markets for coal. Dr. Lawall was made a member of the group, but was unable to attend their meeting in Charleston. He asked me to attend to represent him. There were many prominent people assembled including several coal operators that I knew. At one point, the Chairman called for suggestions from the group. I got to my feet and made three suggestions. The first was that all new state buildings should be heated with coal and a greater effort to be made to improve the highways and secondary roads in the state. I also suggested that a research bureau be added to the School of Mines to devise more uses for coal.Joe Leonard was hired to head up a research effort which was very productive. It was some satisfaction to me that all suggestions were incorporated in their final report. Not long afterwards, a new building was erected at WVU to house the College of Engineering and it was heated by coal.

DIRECTORS

Robert J. Bowman	Detroit
Robert J. Bulkley	Cleveland, Ohio
Barnum L. Colton	Washington, D.C.
Cyrus S. Eaton	Cleveland, Ohio
Cyrus S. Eaton, Jr.	Cleveland, Ohio
Roger H. Ferger	Cincinnati, Ohio
Herbert Fitzpatrick	Huntington, West Virginia
M. S. Fotheringham	Steep Rock Lake, Ontario
Fay A. LeFevre	Cleveland, Ohio
William H. Lipscomb	Leesburg, Virginia
Walter J. Tuohy	Cleveland, Ohio

OFFICERS

Cyrus S. Eaton	*Chairman of the Board*
Walter J. Tuohy	*President*
Robert J. Bowman	*Chairman of the Executive Committee*
M. M. Cronk	*Vice-President-Staff*
John E. Kusik	*Vice-President-Finance*
J. C. Kauffman	*Vice-President and General Counsel*
M. I. Dunn	*Vice President-Operations*
James E. Doyle	*Vice President-Merchandise Traffic*
Charles E. Lawall	*Vice President-Coal Traffic and Development*
A. W. Hix	*Vice President-Purchases and Stores*
R. F. Browne	*Comptroller*
T. H. Keeler	*Secretary*
D. A. Edwards	*Treasurer*

A group picture showing all of the participants at Dr. Lawall's retirement party at the Frederick Hotel in Huntington. Seated left to right; Guy Hoover, Roy Reidinger, Mrs. Lawall, Dr. Lawall, George Sandmann, and J. B. Herring. Standing left to right; Jim Lilly Jim Phillips, Dick Waybright. Earl Kiser, Russell Blue, the author, Preston Henshaw, Jimmy Ward, Bob Leitner, Dick Barnett, Ken Lemons, and Ed Haessler.

1957 - DeVINE COMES ABOARD

The Annual Report started out with the letter from the Chairman and CEO that had been released on January 2, 1958, the first business day of the new year. It was heady news for the shareholders:

"Net income was $68 million. This was $8.36 a common share, compared with last year's all time record $8.28. The annual cash dividend payment attained a new high of $4.00 a common share."

"Coal and coke traffic revenues were a record $233 million. Record tonnages of coal for foreign and domestic consumption were loaded into ships at C&O coal piers."

"Merchandise traffic revenues were just below the record $170 million produced in 1056. In the year, new industrial developments expected to produce $4 million annually in traffic revenues were located on our system."

* * * * *

"National bituminous coal production was good, close to 1956's 500 million tons. This was well above the average for the last half-dozen years."

"C&O loadings were near 1956's 65 million tons, with C&O percentage of national loadings up slightly. This reflected the continuing demand for the high-grade coals originating at mines on the C&O."

"Foreign demand for coal continued heavy. Coal moving abroad over C&O piers was up one-sixth over the year-before record of 20 million tons."

"Merchandise loadings again showed the diversified base of C&O's total traffic. Although a little less than the previous year's record, reflecting general business in the final quarter, the volume of merchandise business as measured in net ton miles was well above the average of the last five years."

* * * * *

"In the year, $88 million in capital expenditures-bringing total expenditures since World War II to $700 million-went into additions and betterments."

"These improvements included a new electronic Car Location Information Center, a bulk materials import pier, a coal loading tower, and an electronic manifest yard."

* * * * *

The report went on to describe more details about the year:

"Suppliers profited by company expenditures of more than $125 million for materials and services."

* * * * *

"The average American, someone with a sharp pencil has figured out, consumes four times as much coal by weight as he does food."

* * * * *

"The average man of the rest of the world, because his life is not so electrified or mechanized as the American's does not yet consume coal in that same four-to-one ratio with food. But for many of these average men abroad, particularly the European, the coal he does consume is greater than his own supply."

"It was this coal hunger, both at home and abroad, that in 1957 kept the American bituminous coal industry busy in its mines."

* * * * *

"To meet this need for coal, total bituminous production in the year was 490 million tons."

* * * * *

"C&O again carried more of this coal away from the mines than any other railroad-its originate coal traffic was 63.7 million tons. This was 13 percent of the nation's production, slightly better than C&O's traditional one-eighth."

* * * * *

"Foreign countries, not including Canada, took a record 58 million tons of United States coal. This was almost 10 million tons more than the year before, the previous record year."

"C&O carried to Tidewater and loaded into ships more of this foreign-bound coal, 23.3 million tons, than any other railroad. It had carried 19.6 million tons the previous year. Aided by a new sixth loading tower, C&O's port at Newport News, Virginia, set a new loading record for both export and domestic coal shipments of 27 million tons."

* * * * *

"It is estimated that by 1980 the electric utilities will require annually 400 million tons, more than double the 1957 consumption."

"Coal's second largest consumer, the steel industry, used 110 million tons of coal in 1957 to produce 113 million tons of steel."

* * * * *

"In 1957, West Virginia's mines reached the point at which they had extracted 6 billion tons of coal from the ground-yet this monumental output had taken only 12 percent of the state's estimated recoverable reserves."

* * * * *

"Developments in 1957, while promising, indicated that the day of nuclear power for all is still far off."

* * * * *

"In 1957, 1.4 million cars loaded with coal were moved over the railroad. Almost an equal number of cars carrying merchandise freight-1.3 million-were also moved."

* * * * *

"Eighty industries, including some of the top corporations, made this decision in 1957 to locate a factory, warehouse, refinery, power plant or other facility on the Chesapeake and Ohio."

* * * * *

"When they all get into business, the 80 new industries are expected to give the railroad traffic that will result in new revenues of $4 million yearly."

"A good example of how industrial development creates C&O traffic is Ford Motor's new Lincoln plant at Wixom, Michigan."

* * * * *

"Pride of the C&O's passenger train fleet is *The George Washington* , overnight train between Washington, D.C. and Cincinnati. The 25th anniversary of *The George,* one of the proud 'name' trains of the country, was observed in 1957."

"It was also a silver anniversary for Chessie, the railroad's sleepy kitty symbol. *The George* was the first air-conditioned long-distance train, and to suggest the comfort of its berth early advertisements pictured Chessie along with the slogan, 'Sleep like a Kitten.' By popular demand, Chessie has been working on the railroad ever since."

"Revenues from passenger service in 1957 were $14.3 million, which was less than 3.4 percent of total operating revenue."

* * * * *

"The price tag on the year's additions and improvements was $88 million, equal to one-fifth of total revenues."

* * * * *

"This two-thirds of a billion dollars spent in a dozen years has made the C&O practically a new railroad."

* * * * *

"Electronic computers are sometimes called 'mechanical brains.' C&O has one of the largest, Univac, in its Finance Department."

"Now both radar and a computer have been teamed up to make C&O railroad operations faster, safer and cheaper. Between them they practically run the new $5.5 million classification yard for merchandise freight cars at Russell, Kentucky, which went into operations early in 1958."

* * * * *

"From the mines C&O coal moves to the Great Lakes as well as to Tidewater. And at C&O's coal port at Presque Isle, near Toledo, Ohio, construction of a new $7 million coal loader was started in 1957. Scheduled for completion in June, 1958, it will be able to load 6,000 tons an hour, three times more than any coal loader ever built. Its revolving car dumpers will be the fastest in the country."

"In 1957 Presque Isle's present loaders set a new record by dumping nearly 19 million tons into vessels destined for Great Lakes ports in the United States and Canada."

* * * * *

"Electronic traffic control has been extended on the system over recent years and in 1957 an additional 100 miles was installed/ C&O is now using this electronic traffic cop to direct its train movements, swiftly and safely, on 1,985 miles of track, including 60 percent of the main line"

* * * * *

"For 55 locomotives, 6,457 coal cars, and 1,148 merchandise cars, the railroad spent $71.4 million in the year."

* * * * *

"Regular maintenance kept 99 out of 100 of all the railroad's freight cars in operating condition in the year, a much higher proportion than the average for the industry."

"Track was kept in excellent shape, too. C&O has no deferred maintenance. New rail was laid on 142 miles of line, and 379,000 cross ties were replaced. Curves were reduced at 30 points between Gladstone and Richmond, Virginia, to speed up traffic on the important James River Subdivision."

* * * * *

"By the end of 1957, employee-owners had acquired a total of 77,477 shares by the payroll deduction plan. This investment of $3.6 million in their company has returned to them more than three-quarters of a million dollars in dividends."

* * * * *

"Gregory S. DeVine, a leading executive of the coal industry with 30 years' experience in coal marketing, joined the railroad as a Vice President in Cleveland in August. Mr. DeVine, age 51, had been Executive Vice President of Peabody Southern Coal Company, of Chicago, and before that was with Truax-Traer Coal Sales for 22 years. He is a graduate of Northwestern University."

"Owen Clarke resigned in January, 1958, as Chairman of the Interstate Commerce Commission, the government body which regulates the railroads, to join the Chesapeake and Ohio as Vice President in Cleveland."

* * * * *

"Mr. Clarke, 44, was Prosecuting Attorney of Yakima County, Washington, and Chairman of the Washington Public Service Commission before being appointed to the I.C.C. by President Eisenhower in 1953. He is a graduate of the Law School of the University of Washington."

"Erle T. Rucker, 50, who joined C&O in the engineering forces in 1928, was appointed Manager of the Northern Region, headquartered at Detroit, in September. He is a graduate of Virginia Military Institute."

"C. Vernon Cowan, 46, who started with the railroad 20 years ago, was appointed Manager of the Eastern Region, headquartered at Richmond, in June. His advancement was through the posts of Terminal Trainmaster, Assistant Division Superintendent and General Superintendent of Transportation."

"Charles E. Lawall retired as a vice president on February 1, 1958, after outstanding contributions to the railroad during an important period of coal growth."

"Robert J. Bowman, who retired in February, 1957, as Chairman of the Executive Committee of the Board of Directors, died on January 22, 1958. In railroading, for over a half-century, Mr. Bowman was President of the company from 1946 to 1948, when he became Chairman of the Executive Committee. Upon Mr. Bowman's retirement, the Board of Directors, in a testimonial, said it was proud to recognize his outstanding record with the Chesapeake and Ohio."

"William H. Lipscomb, a member of the Board of Directors for 15 years, died on July 29, 1957. In a memorial, the Board of Directors gratefully acknowledged his invaluable contributions to the railroad."

* * * * *

"In rating the management abilities of leading railroads, 'Forbes' scored C&O at 95, the highest rating given. Only one other railroad was rated as high."

* * * * *

The arrival of Gregory DeVine on the railroad in August caused a great stir on the entire system. Everyone knew that he was replacing Dr. Lawall and assumed that he would eventually replace the President. He was of the same religious persuasion as the boss and the rumor mill had the story that they were best friends for a long time while Walter was in the coal business in Chicago. Dr. Lawall was kept on until February of 1958 to assist the new man in getting acquainted with his job. Another rumor surfaced in our office from the frequent phone conversations with the office personnel in Cleveland to the effect that Dr. Lawall had persuaded Walter to let him set up a separate office in Huntington to supervise the subsidiary land companies.

When DeVine learned of the plan he went to Walter and said he would resign if the plan was put into effect which quickly ended any such diversion.

A few weeks after Greg's arrival, Dr. Lawall brought him to our Huntington office. Of course, we were all on needles and pins in anticipation of meeting him and more so about whether he would make any changes. We didn't have to wait long. On the second day of his visit, he took over Dr. Lawall's office and called each one of us in to chat. He told me that I would become Administrative Assistant to the Engineer of Coal Properties who was going to be Guy Hoover. My special responsibility was to oversee and motivate the office force and to expedite the mail, especially that which was addressed to the Cleveland office. This recommendation was probably suggested to him by George Sandmann. Evidently our office had become notorious for slow responses because requests had suffered intermidable delays. I was aware of the resistance to pressure for quick recommendations from both Davis and Lawall and was not surprised at the reason behind my instructions. George Sandmann was the number one Assistant in our Cleveland office. He started out as a Clerk in the Huntington office and was promoted to Cleveland before I arrived on the scene.

I knew at the time that my promotion was a heavy blow to J. B. Herring who was much older than I and who could obviously be disappointed that I had been put ahead of him. However, he never said a word about it to me or anyone who might of told me and proved to be a good staff worker and loyal friend in succeeding years.

Guy Hoover was a very fine mining engineer who liked to work on the projects that were the essential elements of our office such as investigating coal properties and making leases. However, he had an introverted personality and no desire whatever to wear the hair shirt and command the troops. Consequently, he was more than glad to have me take over all the internal problems in the office regarding procedures and personnel. This role was right down my alley as I had served as the Executive Officer in my small Navy base outfit overseas and second-in-command at the Materials and Equipment Division at the Interior Department in Washington.

I was determined that our office would continue to perform exemplary work as far as we could under the time constraints imposed by management. I soon implemented our office procedures with methods learned in Washington which required most of the paper shuffling to be taken care of by the clerks. I worked closely with Guy so that he was kept fully informed regarding my adventure and he responded by backing me up when approached about resistance to the new procedures.

My first impression of Greg was very favorable. He was obviously a man of sharp intelligence with a good sense of humor. He appeared self-confident and relished being the boss. He did not hesitate to make decisions.

While DeVine was visiting our office he sauntered into the drafting room where Jim Lilly was working on a map. It was a very warm day and all the windows were up. Greg asked Jim why he had a cloth towel spread out at the bottom of the drawing and he replied that it was to catch the sweat dripping off his chin and arms. This incident obviously made a big impression on the new boss.

A couple of weeks later, Bill Thurston arrived in our office with a mandate to help us fix up the place. He brought an expert who performed similar work in Cleveland and we soon had a plan to open up the office to make a central "bull pen" for the clerks. This move was desirable to bring Jimmy Ward out of his private office where he couldn't hide. He caught on to the game and took early retirement.

The final plans included new window air-conditioners, a three-color decor, new rugs and a fancy sign in the hallway where the elevators were located. The other employees in the building were impressed and envious of our new decor. The only thing comparable in the whole building was the top floor where the new system-wide passenger reservation office was located. It was very plush to accommodate visiting customers.

We immediately started to interview candidates for Jimmy Ward's job. The clerks union had built up several guidelines over the years so that we had to bring in a clerk in a lower category from another department. At Dick Wrybright's urging I hired Linda Ashcraft who was the daughter of another clerk and good friend of Dick's. She was not happy in our group and complained of being cold in the winter. Finally, she decided to quit working to take care of her children.

When Dr. Lawall retired, I knew that I would inherit Russell Blue as my personal secretary and was much pleased at the prospect because he was a good man to have on the job.

The annual meeting in 1957 was a remarkable diversion from past practice by being held in the local Keith Albee Theater in Huntington, with a full house of shareholders. The local press had proclaimed that an important announcement would be made at the meeting of a great boost to the local economy. I had been aware that a large industry such as the railroad would sometimes have a difficult task in maintaining the goodwill of the citizens of a small city such as ours. On this occasion, however, the two parties were entering into a mutual admiration phase.

Mr. Tuohy was at his best in charming the large audience and inspired them by announcing that the headquarters of the two largest departments in the company, Operations and Engineering, would be transferred to Huntington and a new multi-story building would be erected near the Olympic Pool to house the local company offices. My wife and I were in attendance for my only participation in one of these meetings and we could feel the excitement and warm feelings that the news generated.

That afternoon, I was asked to drive Mr. Tuohy to the local Guyan Golf and Country Club where he was to play golf with several local dignitaries. He had been informed that there was a ladies meeting in progress at the club and he questioned me very closely on the names of some of them whom he was likely to encounter. From this experience, I learned why he was so successful in interpersonal relationships. In retrospect, it is obvious that Messers Fitzpatrick and Dunn had engineered the move to the city. "Mr. Fitz" had advocated moving the company headquarters to Huntington for many years due to its location near the center of the rail system. Mr. Dunn joined him in the effort because it would reduce his travel time considerably.

☆

1958 - A DOWN YEAR FOR BUSINESS

The annual report merely repeated the Chairman and President's report issued in the Flash Annual Report sent out on January 2, 1959. An unusual feature of the report was a biographical sketch about the Chairman, President and the new Vice President of the Coal Department which are quoted as follows:

"Cyrus Eaton has been a Director of Chesapeake and Ohio for 16 years and Chairman of the Board for five years. His other business activities include the coal and iron mining, steel and electric power industries. In his non-business affiliations, Mr. Eaton is a Trustee of the University of Chicago, Case Institute of Technology and Denison University, an Elector of the Hall of Fame at New York University and a fellow of the American Academy of Arts and Sciences. A native of Pugwash, Nova Scotia, he is a graduate of McMaster University, Toronto."

"Walter J. Tuohy has been President and a Director of Chesapeake and Ohio for ten years. He came to the railroad as Vice President of Coal Traffic and Development in 1943, after 20 years in the coal business. He is a Director of the Association of American Railroads, Cleveland-Cliffs Iron Company, American Coal Shipping, Inc., and the American Overseas Investing Company. He is a graduate of DePaul University in Chicago, where he was born."

"When Gregory DeVine came to the C&O in 1957, he brought to the dynamic business of selling transportation more than thirty years in the equally swift-paced business of selling coal. The combination of experience is one he shares with C&O's Chief, Walter Tuohy, himself a coal sales executive before he entered railroading."

"Mr. Devine has been in the coal business since 1926 when he made his first calls as a sales trainee. In 1934, after service with two other companies, he joined the Binkley Coal Company in Chicago as a salesman. Seventeen years later he was its President."

"Binkley in 1952 became a part of Truax-Traer, the big coal mining and marketing firm whose West Virginia properties are served by C&O. Gregory DeVine became Vice President of Truax-Traer and, later, Executive Vice President of Peabody Southern Coal Company. From the latter position, he came to C&O."

"C&O's coal sales chief was born in Rahway, New Jersey. In his youth his family moved to Chicago where Mr. DeVine graduated from St. Patrick's High School. He attended St. Viator's College in Bourbonnais, Illinois, and Northwestern University."

"Mr. DeVine and his wife, Ursula, have three children: Barry, Jon and Suzy. They live in Lakewood, Ohio."

The annual report started out with a table of highlights comparing results in 1958 with those of the previous year.

	1958	1957
Dividend Paid per Common Share	$4.00	$4.00
Earned per Common Share	6.36	8.36
Operating Revenues (millions)		
Coal and Coke	177	233
Merchandise	151	168
Other	28	31
Total Operating Revenues	356	432
Expenses, Taxes, etc.-Net	304	364
Net Income (millions)	$52	$68
Net Income % of Operating Revenues	14.5	15.6

There was a new face listed with the Board of Directors, Erwin H. Will, who was elected on January 27, 1959. He was Vice Chairman of the Board of Virginia Electric and Power Company. This company was the largest utility on the eastern seaboard and he was obviously elected to promote more business for C&O service. The report continued with the following elaboration:

"National production of bituminous coal was down about 90 million tons from 1957's 490 million. Export coal was off as a result of business and seasonal conditions in Western Europe. C&O's coal and coke revenues still compared favorably with the average revenues of the last ten years."

* * * * *

"Merchandise freight revenues, lower by $17 million than in 1957, reflected the pace of business. New sources of C&O traffic continued to develop: a diversified group of industries chose locations in C&O territory for 90 new plants, expected to produce $4 million more in annual revenues. Capital expenditures were about $21 million, compared with $88 million the year before."

* * * * *

"At 1958's end, working capital was about $55 million, compared with $44 million a year ago."

* * * * *

"A $7 million facility for loading coal from rail cars into ships was placed in service at C&O's Lake Erie docks near Toledo. Electronic traffic direction systems were extended to additional areas of C&O's main lines. Five hundred new coal and merchandise cars were added to the company's fleet."

* * * * *

The remainder of the report was devoted to a four page dissertation by Greg DeVine about coal traffic. The gist of his story follows:

"The nation's coal production was down in 1958 in line with general business conditions. For example, the electric utilities, coal's largest customer, used five million tons less than they did in 1957."

Gregory DeVine presents Guy Hoover with his retirement booklet as the author, Fred Toothman looks on. 1960

"But the sharpest drop in coal use came from the steel industry. It consumed only 83 million tons, compared with 110 million tons in 1957."

"Export traffic was off, too, but for reasons somewhat different from those which affected domestic use. In 1956, during the Suez Canal crisis, European customers bought more coal than they could use right away because they feared that their supply lines for oil would be cut off."

"On top of that, Europe had mild winters in 1956 and 1957, which meant that those huge stockpiles weren't cut into very deeply. And finally, West Europe's steel industry experienced a slight recession of its own in 1958."

"As a result, all export loadings from the United States dropped off to 38 million tons, compared with nearly 58 million tons in 1957. C&O's loadings were 15 million tons as against 23 million tons in 1957."

* * * * *

"While I don't feel we can expect, in the years just ahead of us, to achieve the same records we did in 1957, I do think the outlook is good."

"The electric power and the steel industries will take the greatest share of the bigger output from the mines in 1959. The utilities will use 163 million tons--more than they have ever used before."

* * * * *

"The long-range future for steel, however, can only be described as spectacular. Leading authorities predict that by 1975 the industry will be able to produce 200 million tons of steel annually."

"As for export coal traffic--it is not going to climb back to its recent high levels. At the same time, I am optimistic about the overall future of the export market."

* * * * *

"Today, C&O serves a total of approximately 300 mines and the area continues to attract new producers. In 1958, six new mines were opened, one of which represents an investment of $5 million by its owners. Several more mines resumed production after temporary shutdowns."

* * * * *

"There is one more factor in coal's future that we ought to discuss. I am asked frequently what effect I think nuclear energy will have on coal's position as the leading fuel for electric power."

* * * * *

"It is so expensive that the United States government must underwrite most of the cost of building and operating a nuclear power plant. Those costs are several times greater than the costs for a conventional coal-using plant."

* * * * *

"Certainly, the use of nuclear energy will increase. But that increase will be very small compared to the tremendous increase in the use of coal as the utilities' basic fuel."

* * * * *

"When I was in the coal industry, C&O was always looked on as the railroad that did the best service job. We always felt safe with C&O because we always got our coal when we wanted it."

* * * * *

"To improve our service to customers, we are applying C&O's Car Location Information Center and its system-wide communications network to the movement of coal traffic. This will give us an even better car reporting system than we have now."

* * * * *

The huge amount of ink devoted to DeVine's plugging of coal indicated that there was little doubt that he was being groomed as the successor to President Tuohy.

When the annual coal traffic meeting was held at the Greenbrier in 1958, Greg DeVine made it clear to all of us that he was in charge and there would be some changes made in our operations. Specifically, he let it be known that he did not approve of our customary long drawn-out staff meeting programs and called off any post-dinner sessions. When a couple of the agency heads got up and complained about directions emanating from headquarters, he let them know in no uncertain terms, that such discussion was out of bounds and should be brought to him in private.

We also learned that he liked to "party" and had a lively wit and convivial personality. He kept his table entertained at the Old White with stories and comments interspersed with frequent rounds of libations. His own preference was martinis and he could consume more with less inebriating effect than anyone I had ever encountered. At that time, West Virginia State Law required that each guest who ordered alcoholic drinks must have his own supply purchased beforehand and stored in the Club. Greg made sure that his supply was ample and he was very generous in sharing with guests.

I noticed that Greg always had a retinue of men who accompanied him while walking around in the hotel. The man who was Chief Engineer on the staff was usually included. I also noticed that whenever he received any kind of service or made a request of any member of the hotel staff, he slipped a $10 note into the individual's hand. After watching our new leader, I became much more generous with tipping and recorded them on my expense account.

A sad note was received by all C&O personnel when the national press announced the death of Robert Young at his home in New England. It was surmised that he had become disenchanted with his hope that the New York Central would make big money and he felt responsible for misleading his Texas friends in a losing effort. Accounts of his death included the grisly detail that he had inserted the barrel of a shotgun in his mouth and blew his brains out.

In the spring of the year, Greg notified Henshaw and I that he wanted to make a four day trip to the New River Field where our competition from the N&W and Virginia Railway was most severe. We stayed in the Beckley Hotel and ate our dinners at the Black Knight Country Club. Greg was most interested in meeting the operators in the field and spent a great deal of time with Alex Hunt who was the power behind the New River Company. I greatly admired his knowledge and learned a great deal from our association over the years in my job. The railroad was much interested in the company because of the large tonnage it represented in traffic from jointly served mines. E. V. Bowman was a mining engineer about my age who was President of New River, and was a very intelligent coal operating manager.

During the year, we learned that Greg had to contend with claustrophobia and felt trapped in riding his office car. He even resorted to riding on the outside platform to ease his distress. Riding in airplanes did not seem to bother him unless the flight was very long. He let us all know that we should use commercial flights in our travels. I learned that a cocktail before and enroute eased my apprehension about flying.

The plan for constructing a new office building in Huntington was dropped when it was decided to purchase a war surplus building in the west end which had been a plant operated by Sylvania Electric to produce miniature electronic vacuum tubes for the proximity fuses in anti-aircraft shells. The entire interior of the building was removed and replaced by offices for the operating department, the engineering and mechanical departments, as well a real estate. It was later called the OH building.

The local company employees were titillated by the news that M. I. Dunn had a fist fight with Basil Brangham in his office car. The apparent cause was disagreement over terms of a new contract for the Clerks Union of which Brangham was Chairman. It was reported that "M. I." suffered a black eye and the interior of the office car was badly damaged in the fracas. Both men were powerfully built so that it must have been a real donnybrook.

When Greg joined our department in Cleveland, he was assigned Ted Boczek as his secretary who had previously worked for Dr. Lawall in the Cleveland office. Greg promised Ted that if he stayed with him for one year, he would be promoted into a management position.

As a member of the same church, I met J. D. Francis several times and later attended his funeral which was the largest one I had ever attended. He was the man who developed Island Creek Coal Company into national prominence.

☆

1959 - TUOHY SIDE LINED BY HEART ATTACK

As had become the custom, the results were published in a Flash Annual Report on the first working day of the new year in the form of a letter signed by the President and the Chairman:

"At the start of the new year--the 175th anniversary of C&O growth and progress--this report is sent to you with pleasure."

* * * * *

"The nation's business came back strongly, as predicted, form the lower levels of 1958. However, the pace slackened as the steel strike, beginning July 15, stretched out for 116 days, longest interruption of steel production in history."

"The brisk pick-up in C&O traffic which began late in 1958 was reflected in an excellent first half-year. Even when the steel strike came at mid-year, C&O's revenues and earnings held up well. For the full year, transportation revenues surpassed those for 1958, except for export coal."

* * * * *

"Merchandise traffic. Revenues from the movement of merchandise were $10 million greater than in 1958. Demonstrating C&O's traffic diversification, the revenues from moving merchandise were practically the same as the revenues from moving coal. Eighty industries chose sites in C&O territory for new plants, expected to furnish an additional $2 million in annual revenues."

"Coal Traffic. A considerable step-up in orders from coal users in the United States and Canada, mostly before the steel strike, gave C&O a $10 million increase over 1958 in non-export traffic revenues. Reflecting continuation of the business and seasonal conditions that have reduced coal consumption in Western Europe, C&O's overseas export coal traffic was down. National cost production about equaled the 410 million tons of 1958."

"Innovation and Progress. Piggyback, the carrying of highway trailers on flat cars, produced encouraging results. To offer shippers more flexible schedules and develop additional income from passenger operations, C&O operated piggyback cars on passenger trains. Railvan, the unique freight-carrying trailer developed by C&O research to operate on either highways or rails, went into daily test service in Michigan."

"Finance. C&O ended the year stronger, financially and physically, than at any time in its long history. Working capital at year's end stood at its highest level, $61 million. Because of C&O's physical soundness, the result of an improvement program costing more than $700 million since World War II, capital expenditures of only $16 million were necessary in 1959"

* * * * *

Highlights	1959
Dividend Paid per Common Share	$4.00
Earned per Common Share	5.60
Operating Revenues (millions)	
Coal and Coke	162
Merchandise	161
Other	25

Total Operating Revenues	348
Expenses, Taxes, etc.-Net	302
Net Income (millions)	46
Working Capital at Year End (millions)	61

"A good year is predicted for business generally in 1960. This favorable outlook, of course, is predicted on there being no major work interruption."

* * * * *

The rest of the annual report was in the form of a discourse by each of three Vice Presidents; Doyle, DeVine and Dunn as well as brief statements about Public Relations, President's staff, Research and Advertising.

James E. Doyle was described in a brief biographical sketch and his abstracted presentation followed:

"A Virginian, whose father built locomotives for C&O, Jimmie Doyle went to work as an office boy for the railroad 39 years ago in Richmond, started learning the traffic business while attending night classes at Virginia Mechanics Institute and William and Mary Extension. His career in traffic sales and service brought him to Cleveland in 1929 and to his present post in 1956."

* * * * *

"Our revenues of $161 million were six percent greater than in 1958. When I reported to you last year, I told you that the outlook for 1959 promised higher revenues and carloadings. If steel production had not been interrupted for the long period of 116 days, revenues and carloadings would have been much higher than they were."

"As it developed, despite adversities, 1959 was our fourth best merchandise year. And from all indications now, we will surpass the 1959 figures in 1960."

"The year was successful not just from a revenue standpoint; 1959 stands out for the new things C&O was able to do. We began operating trailer-on-flatcar, popularly called 'piggyback,' service, and quite successfully, too."

* * * * *

Gregory S. DeVine expounded on the Coal Department and his abstracted statement follows:

"There was a slight increase in the use of coal in this country and Canada, despite a 116-day steel strike. It was created largely by the expansion of the electric power industry."

* * * * *

"The result for C&O: revenues from the transport of non-export coal of $124.6 million, a boost over 1958 of $9.6 million."

"Very important, too, for C&O's coal business was the permission from the Interstate Commerce Commission to offer coal users for the first times a new kind of transportation pricing system called volume rates."

* * * * *

"In the particular case where the ICC made its historic ruling, the new rates retain for C&O and co-operating railroads about $10.5 million in annual revenues."

"In another step toward effective competition, C&O sought and received the ICC's permission to offer an attractive new rate for all-rail movement of coal to a major steel company. This traffic, about 1.4 million tons a year, moved principally in the steel company's barges. Now it will be transported by rail."

"There was one other event in the year that promises much for the coal industry and the railroads which serve it. This was the formation of the National Coal Policy Conference, an organization of coal-carrying railroads, coal producers, the United Mine Workers, coal equipment manufacturers and coal-burning electric utility companies."

* * * * *

"For the nation as a whole, overseas exports of coal were at the lowest level since 1954. Only 24.9 million tons from American ports, 9.1 million tons of it handled by C&O. In 1958, the national export total was 38.2 million tons, with C&O's share 15.3 million tons."

* * * * *

"Coal from this country is of a better grade, particularly for steel-making, than anything Europe has. It can be mined faster and cheaper here."

* * * * *

"Some foreign countries which were the biggest users of high-quality American coal have either placed prohibitively high tariffs or low import quotas on all foreign coal."

* * * * *

"The electric power industry used 170 million tons of coal in 1959. In 1960, it estimates it will burn 190 million tons."

"In 1960, the steel industry expects to produce from 120 million to 125 million tons. The record production was 117 million tons in 1955. Since it takes about three-quarters of a ton of coal to produce a ton of steel, the bigger output will mean a bigger market for coal."

"To supply these big customers and other industries, the nation's coal mines will produce about 436 million tons in 1960, up from 400 million tons in 1959. C&O's carloads should go up, as a result, about 7 percent."

* * * * *

"Recently we received a letter from an industry which uses C&O coal. It had bought an older plant and was faced with the prospect of replacing its coal-fired boilers at considerable expense."

"At that point, one of our fuel service engineers offered his counsel: use the same equipment with some modification and continue the use of coal. The company said: 'His enthusiasm and knowledge made it a pleasure to work with him, and the C&O has certainly rendered us a very valuable service by making this assistance available. We fully realize that his recommendations have saved us a great deal of money.'"

* * * * *

M. I. Dunn was described in a brief biographical sketch and it follows along with his discussion about the Operating Department:

"A third generation C&O'er, whose grandfather helped rebuild and became Chief Operating Officer of the line after the War between the States, Mr. Dunn has been railroading for 43 years. He became Vice President for Operations, in charge of the Transportation, Engineering and Mechanical Departments, in 1956. Born in Virginia, he was graduated from Washington and Lee University and holds an Honorary Doctorate from his alma mater."

* * * * *

"Although C&O's gross revenues declined, our road and equipment maintenance program was carries out in full. No deferred maintenance was accrued."

"Only 2.2 percent of C&O's fleet of 96,000 freight cars were in need of repairs at year's end, down from the ratio of 5.2 percent in January, 1959, and a considerably lower ratio than the average for the entire industry."

"More rail, ties and ballast were applied to make your railroad a stringer, better-maintained property."

"However, while these programs were successfully accomplished, payrolls exceeded those of 1958 by only one million dollars, or 6/10th of one percent."

* * * * *

"During the year, we purchased 500 box cars equipped with special loading devices, and we began converting 125 standard cars to specially-equipped units in our shops. We acquired 150 flat cars, each long enough to carry two highway trailers, for our piggyback services. Our car rebuilding and conversion program will continue through 1960."

"In the past, C&O's seven-ship fleet of Lake Michigan trainferries has been operating two terminals at Milwaukee, Wisconsin. In 1959, we were able to conclude negotiations for the consolidation of all our Milwaukee operations at the Jones Island Terminal. Completion of this project--approximately June 1, 1960,--will increase utilization of our fleet equivalent to the addition of another vessel like the Spartan or Badger, built in 1952 at a cost of $5 million each."

* * * * *

John S. Marshall wrote about the Better Service Conferences:

"At important points on C&O, there are 23 Employees' Better Service Conferences. Their monthly meetings serve further to unite the thinking and loyalty of thousands of C&O people toward the railroad's goal of doing the best job possible for its customers."

"Many C&O people possess an additional stake in the company's future in their roles as shareowners. At the end of 1959, they held 63,485 shares of C&O common stock as members of the Employees' Stock Purchase Plan and received during the year about $250,000 in dividends."

* * * * *

"To coordinate these company-wide manpower planning activities, the President's Personnel Conference was established early in 1959. Composed of representatives from all departments, the Conference provides a new opportunity to exchange experience and coordinate plans to mutual advantage."

* * * * *

Kenneth A. Browne presented a description of usable projects:

"Railvan, the freight vehicle C&O developed to operate either as a highway trailer or on rails as part of a train, was placed in daily experimental service in 1959. Seven units are operating on the Northern Region, carrying mail and express between Grand Rapids and Traverse City and Grand Rapids and Detroit."

* * * * *

"The Research Department also conducted important studies for other departments. For example, tests proved the effectiveness of new devices that protect merchandise from damage while it is being shipped. A new journal bearing that will help to eliminate 'hot boxes' was designed and is being fully developed by a railway supply manufacturer under license from C&O."

* * * * *

Walter S. Jackson commented on the advertising campaign that year:

"Special series of advertisements in leading magazines, newspapers and the financial press highlighted C&O's export and import facilities at Newport News; the benefits to shippers from our Car Location Information Center; the better, faster east-west routing available through C&O's Lake Michigan trainferries, and the excellent sites for new industries in the territory C&O serves. Other campaigns featured coal transportation and the attractions of C&O's passenger service."

* * * * *

Two of the most important events affecting the C&O in 1959 were not even mentioned. The first of these was a tragic blow to the management when Walter Tuohy came down with a heart attack and was out of action for several weeks.

While he was recuperating, the N&W Railroad President, Stuart Sanders, notified our management that he had negotiated a merger agreement with the Virginian Railway and requested a meeting to discuss this matter. It was obvious that these negotiations were conducted in complete secrecy and must have caught our company by surprise. Since Mr. Tuohy was out of pocket for the moment, Eaton and DeVine met with him to learn that several concessions would be made to C&O for agreeing to withhold opposition to the merger in the hearings before the ICC.

The major concession to C&O was to agree to permit routing our trains over the Virginian line from Mullens to Gilbert, West Virginia, thereby permitting C&O coal moving to western markets a shorter route to Huntington via our Logan branch line. They also agreed that C&O could solicit coal traffic from the Virginian local mines on this line. Experience later proved that arrangement for a pilot on C&O crew and other restrictions made the new route undesirable. Not one car of coal traffic was ever obtained from any of the local Virginian mines and our management had to conclude that N&W

Left to right: Joe Daugherty, Speed Gray, Earl Kiser, Arthur Maynard, John Kusik, Fred Toothman, Unknown Coal Buyer and Island Creek Official had elicited iron bound guarantees from all of them that no traffic would be diverted. The one good thing to come out of this experience was the fact that C&O caught the merger fever from this experience which was later to prove so successful.

My own career took on an added responsibility when my wife and I decided to build a new house on East Inwood Drive on the hills behind the city of Huntington. Guy Hoover graciously offered to let me take time off when necessary to go to the project to supervise in lieu of my regular vacation time. We contracted other services such as wiring and plumbing. After three months of hard work, we moved in on the first of October with a lot of clean-up and finishing still to be done.

Since Guy Hoover was approaching retirement age in 1960, I obtained permission to hire another mining engineer to take his place. At an AIME meeting that spring, I asked several leaders in the group for suggestions. Joe Malesky, who worked for the Bureau of Mines in Norton, Virginia, suggested to me that one of his inspectors wanted to move back into industry. His name was Frank Gaddy. He proved to be a good candidate who was hired on June 1, 1959. He was a graduate of VPI with a wife and two children.

During the year, Greg DeVine had advised me to work with the President's staff in Cleveland who were pursuing Tuohy's instruction to investigate investment in the petroleum business to protect the company's supply of diesel fuel. Rudnicki was following up on this and convinced them to buy into a secondary oil field recovery project located near Ashland, Ohio, and on December 1, he was transferred to the Cleveland office.

The year was a busy one for changing personnel in our Huntington office. Greg soon learned that we had lots of problems with our side-tracks. In many cases, operators had made agreements which permitted them to use the tracks but were later forced out of business but continued to hold onto the track for renting to other operators or in the hope of resuming their own operations. It was to our best interest to free up the idle tracks and seek new business for them. To perform the paperwork in these matters, Ed Haessler was put in charge and authorized to hire his own secretary. Fran Huddleston was transferred from another office and proved most efficient in helping Ed work on these agreements.

In the early winter of the year the County Commission of Raleigh County started to petition the C&O top management to assist them by selling a tract of land right in the middle of our Newberry Lands to build a dam and lake impoundment for a recreation park. We were aware of this promotion and learned that this project would directly interfere in the future development of a coal mine for which purpose the lands were intended. We reported these facts to DeVine who replied that he did not want to be "a dog in the manger" but was depending on me to dissuade the idea. This directive really put me on the spot and I had to come up with a solution. The only possible approach I could think of was maybe we could persuade them to relocate to another site. We pulled out all of our maps and studied them intensely finally deciding that Stephens Branch on the edge of our lands might be a good alternative. Our drilling records showed that the coal pinched out on that side of our property. I made a fast trip to the location and decided we might be able to persuade a change to this site. It had a redeeming feature in that there was a grove of cedar trees and another of pine which dressed up the area compared to all the bare trees denuded by winter weather.

A meeting on the ground was arranged to show the area to the County Commissioner. In the meantime I had received permission to offer them 100 acres of our surface land near the area as a donation. The green foliage really impressed them and, a short time later, they agreed to move the project. This was a big feather in my cap to impress upon management my ability to handle my new responsibilities.

On my visits to the Greenbrier during the past year, I had noticed that a new wing was being constructed right on the edge of Route 60 which is the main access highway and to the hotel and the small town of White Sulphur Springs. The announced purpose of the new wing was to house the clinic which was outgrowing its old quarters in the hotel. During the construction, a large pile of earth and rock was being deposited on the other side of Route 60. I suspected that a much larger excavation than just a basement was underway.

When the new wing was opened for use the next year, two new elevators had been installed to serve the upper five floors and the basement. One of the elevators had an extra panel of operating buttons which required keys to operate. It then became obvious to me that there was a large underground part of this building intended for some classified purpose. There was a lot of talk at this time about bomb shelters because of the cold war and I really assumed that this was a special one for some important people.

While this book is being written, an article has appeared in the New York Times of May 30, 1992, blowing the cover for this facility which renders it useless for the intended purpose:

"When it was reported this week that the Federal Government had a bunker at a luxury resort in West Virginia to which members of Congress would be evacuated in a nuclear war, the lawmakers had more to fear than the disclosure of a closely guarded national security secret. There were fears of political fallout as well."

* * * * *

"The articles describe the bunker as self-sufficient in water, electricity and sewage treatment for up to 800 people, and equipped with banks of computers."

* * * * *

"Since the shelter could not stand a direct hit from today's more sophisticated weapons, its disclosure seems likely to doom it as one of the Government's principal evacuation sites."

* * * * *

"The cost of the bunker, built under the cover of adding a new wing to the Greenbrier, is not known. The cost of maintenance is $50,000 to $60,000 a year, according to *The Post*: Ted Kleisner, President and Managing Director of the resort, owned by the CSX Corporation, did not dispute that assertion in an interview today."

* * * * *

"He said that while only a few members of Congress knew of the bunker, about 100 of the Greenbrier's carpenters, electricians and other staff were familiar with it in the course of their duties."

* * * * *

"The resort's statement said it was 'proud to have been a part of the nation's effort to preserve Constitutional government in times of great danger to the nation."

* * * * *

Speed Gray told me that Greg informed all his assistants that Ted Boczek was available for promotion. Speed says he realized that Ted was a good man and took the opportunity to help both men by requesting that Ted be assigned to him as his assistant. This arrangement worked out very well over the ensuing years. Greg reassigned a girl from the office by the name of Louise Fratesi who was a sharp individual to be his secretary. Louise came to the C&O after working for Pocahontas Fuel in Cleveland. She was always very friendly and helpful and I was pleasantly surprised in later years to hear that she and Ted had gotten married.

Greg had a good sense of humor and he often told me the story about Speed Gray and the "office car caper." Speed loved more than any other part of his job to take trips with a customer on Office Car 21, which was a very plush office car assigned to the Coal Department. Speed came into Greg's office and asked him if he could take his friend George on the office car. George worked for a big utility customer on C&O and wanted to go to Newport News to check on one of their shipments. Greg told Speed it was okay and to go ahead. That evening as Greg was leaving the office he passed by the open door to Speed's office. Speed was on the telephone and just as Greg went by he heard him say, "But George, you gotta go, the boss said we could use the office car."

Greg also liked to tell the story about the man who went bankrupt with a new company to make dog food. He built a very modern plant, hired the best people and insisted on the finest ingredients and the best packaging. There was only one thing wrong, the dogs didn't like it.

On another occasion when Speed was riding the office car, he suffered a heart attack right after the train left Charleston. The conductor made an emergency train stop in Montgomery and he was rushed to the Laird

Memorial Hospital where Dr. Laird held forth. During his recovery, Speed became great friends with the doctor and developed a craving for Ballards hot sausage. After Speed got back to Cleveland he asked me several times to bring him some of his favorite sausage which was only available here in southern West Virginia. Finally, I obtained a plastic foam container and took him a couple of pounds of his favorite breakfast meal on several of my trips to headquarters. He always thanked me profusely and one time he said that the sausage didn't mean as much as the fact that it meant I was thinking of him. His heart kept giving him problems after his first attack and finally forced him to take retirement and he didn't live long afterwards.

Speed Gray was in his glory whenever he could entertain a table full of customers in the Old White Club at the Greenbrier. He was a master at telling jokes and could keep up the game for an hour at a time. The way he operated was to order a new drink and excuse himself for a trip to the men's room. While performing the stand-up relief in the toilet, he got out his little black book and reviewed another page of reminders of several jokes. He then returned to the table and went through the jokes he had reviewed. This went on for some time in like manner.

One of his favorite jokes was about the wealthy man who gave his hired hand a pint of home brew for a Christmas present. After several days went by, the man asked him how he liked his present. "That present was just right," said the hired hand. After recalling this incident several days later, the giver asked the recipient what he meant by "just right." His answer was quickly stated, "Why it's as plain as the nose on your face; if it had been any better, you wouldn't have given it to me, and if it had been any worse, I couldn't drink it."

Dick Waybright reached retirement age and seemed ready to give up the daily grind but it was obvious that he had mixed feelings about leaving. I was forced to find a successor in the job of chief clerk which was an important job to keep a smooth running office force. Kenny Lemons wanted the job but I decided that Robert Leitner was the more likely candidate. Kenny went to Hoover and complained that he had been unfairly passed over since he had been in the office longer. Hoover agreed with Kenny even though he had agreed to Leitner's appointment beforehand. This left Kenny completely confused and he seemed to take the change in stride.

Barnett and Leitner were both later promoted out of the clerical ranks to become field representatives and Kenny finally got the job of chief clerk.

☆

With the old B&O passenger station in the foreground, the Terminal Tower dominated the former railroad complex in downtown Cleveland in bygone days. Now, new skyscrapers compete but none cane outshine the Tower.

M. I. Dunn, *Vice President of Operations*

1960 - MERGER FEVER TAKES OVER

The Annual Report opened with a direct appeal to the stockholders to support a proposed merger with the B&O. It was based on the following objectives quoted:

"To the railroad industry in the East, a combined B&O-C&O would bring a vitality unknown for many years. During affiliation, the C&O would assist its partner in restoring its strength and earning power, benefiting almost immediately that part of the public which relies directly on the B&O for its transportation requirements."

* * * * *

"B&O and C&O combined would create a railroad with more than 11,000 miles of line and, based on 1960 figures, estimated annual operating revenues of $732 million and total assets of $2.3 billion."

"Both railroads have achieved a healthy diversification between merchandise and coal traffic, and in types of coal, they are not competitive, but complimentary."

* * * * *

"Each railroad serves important market areas the other does not; each can offer shorter routes between major market centers."

* * * * *

The actual plan for the merger were spelled out in the following two pages of the report which are excerpted as follows:

"The B&O-C&O story began on May 18, 1960, when Howard E. Simpson, President of the B&O, and Walter J. Tuohy, President of the C&O, jointly announced that C&O had made an offer to exchange shares of C&O common stock for shares of B&O common stock 'as a first step toward merger of the two properties."

"C&O's offer is to exchange one share of C&O common stock for one and three-quarter shares of B&O common, and one share of C&O common for one share of B&O preferred."

"On September 14, at a special C&O shareholders' meeting in Cleveland, the Exchange Offer was approved by over 99 percent of the shares represented. Of the 8.2 million C&O shares outstanding, over 81 percent were voted."

"When the Exchange Offer terminated on December 14, 60 percent of B&O's shareowners, holding 55 percent of B&O's total stock, had accepted."

"On December 22, C&O's board, in the light of the majority of B&O shares obtained and the numerous requests from B&O shareowners, announced that C&O would start January 1, 1961, to accrue for B&O shareowners the difference between B&O's rate of dividend and C&O's rate of dividend based on the exchange ration, 'so that assenting B&O shareowners may begin enjoying some of the prospective benefits of being C&O shareowners."

* * * * *

"To initiate coordinated operations in the interests of effective public service at the earliest possible date, the statement said, C&O will make available to B&O up to 5,000 coal cars and up to 100 diesel locomotives. The B&O and C&O, it was stated, 'will immediately enter upon joint studies for further operating and financial assistance."

"After success of the Exchange Offer, it was stated, C&O will use its best efforts 'to consummate a merger of the two companies as soon as practicable upon the following terms:"

"(1) The exchange ratio shall be the same as in the Exchange Offer;

"(2) The President and all members at that time of The Baltimore and Ohio Railroad Company Board shall become members of the Board of Directors of the merged company together with an equal number of directod from The Chesapeake and Ohio Railway Company;

"(3) The name of the merged company shall include 'Baltimore.'

"(4) A major general office of the merged company shall be maintained in Baltimore, with an office of the President, and

"(5) The present B&O pension plan shall remain in effect after the merger for all B&O employees covered thereby on the merger date."

* * * * *

"The Interstate Commerce Commission will set hearings on C&O's application for approval of its Exchange Offer and for authorization to affiliate with B&O."

* * * * *

The usual report issued as a letter from the Chairman and President on the first day of the year contained the following pertinent information:

"Your dividends are your share of the profits C&O made in 1960, despite the fact that business was no better and in some cases not as good as it was in slow-moving 1959. The highlights of the year's results in traffic, revenues and income are in the table."

Highlights	1960
Dividend Paid per Common Share	$4.00
Earned per Common Share	5.15
Operating Revenues (millions)	
Coal and Coke	157
Merchandise	162
Other	25
Total Operating Revenues	344
Expenses, Taxes, etc.-Net	302
Net Income (millions)	42
Working Capital at Year End (millions)	63

"As it enters 1961, C&O's financial strength is greater than ever. Working capital of $63 million at the end of 1960 set a new record for the third straight year. Cash and short-term investments totaled $66 million. Continuance of C&O's maintenance and improvement programs kept your railroad physically strong, too."

* * * * *

"Merchandise traffic revenues increased slightly over 1959's $161 million, even though business conditions were generally unfavorable."

* * * * *

"C&O's new piggyback operations did well. Piggyback was extended to include practically the entire C&O and the number of routes was tripled over 1959. Late in the year, we began transporting automobiles on bi-level and tri-level flat cars. From the 66 new industries which selected building sites in C&O territory, we anticipate potential additional revenues of $2 million annually."

"Revenues from coal traffic were $157 million, compared with $162 million in 1959. Industry's slower pace softened coal's American and Canadian markets and the depressed European market for coal did not improve."

* * * * *

"C&O did housekeeping in 1960, preparing for the expected pickup in 1961. To maintain equipment and roadway, C&O spent $4 million more than it did in 1959."

* * * * *

"On total operating revenues of $344 million, C&O's net income was $42 million, or $5.15 a share. Even in a year of sluggish business activity and higher costs, C&O was able to carry through to net income more than 12 percent of its gross revenues."

* * * * *

The rest of the report was a rundown by the three major department Vice Presidents of Merchandise, Coal and Operations:

"In 1960, our best business came from the movement of such commodities as paper, machinery, auto parts and chemicals, for which shippers are increasingly demanding special handling. Those who can provide it are the ones who get the business."

* * * * *

"The established patrons like Armco were joined by other industries--66 of them, to be exact--who picked C&O-served locations from Newport News to northern Michigan for new plants and other facilities. You will recognize all of these names: Continental Can, A&P, United States Plywood, John Morrell & Company, American Can, General Foods, Pittsburgh Chemical, Johns-Manville,"

* * * * *

"America's bituminous mines turned out in 1960 about the same tonnage they did in 1959. The increase in the use of coal by electric utilities was offset by the low operating rate of the steel industry."

* * * * *

"C&O's optimism about the outlook for coal is based on solid signs of growth we could see in our own territory during the past 12 months."

* * * * *

"You have read much, I am sure, about the excellent prospects for the continued expansion of the electric power industry. What you have read is true, believe me. We saw a good example of it last summer right in our own front yard. Near Louisa, Kentucky, on the Big Sandy River, the Kentucky Power Company started construction of a $39 million electric generating plant."

* * * * *

"Of importance to us and to you is the fact that at least 80 percent of the coal consumed by this new plant will be transported by C&O, under the terms of a new contract rate we established with Kentucky Power. The first of its kind for coal to be set by a railroad in this country, it provides for a reduction under the normal transportation rate on coal in return for a guaranteed percentage of the tonnage shipped."

* * * * *

"In 1960, C&O took these further steps to protect its present good position and equip itself to obtain a larger share of the highly competitive transportation market:

—Spent more than $7 million for physical additions and betterments.

— Replaced 22,500 tons of rail and more than 200,000 crossties.

— Placed under Centralized Traffic Control sixty-eight more route miles, or 109 track miles. C&O's trains are now operated by CTC over nearly 50 percent of the system's 5, 100 route miles.

— Built in its own shops several hundred special merchandise freight cars, including jumbo-sized covered hopper cars, insulated box cars and special 'compartmentizer' box cars.

—Joined 22 other railroads as owners of Trailer Train Company, gaining access to the largest piggyback fleet in the world.

— For our freight car fleet, at 97,000 units one of the largest and most modern in the world, rebuilt in our own shops more than 1,000 coal cars. In 1961 we plan to rebuild 2,000 more hopper cars."

* * * * *

The Baltimore and Ohio Railroad had started providing transportation service between the Ohio River and the Eastern Seaboard exactly twenty years before the C&O. Professor Charles Henry Ambler described this early history of the pioneer railroad in his West Virginia studies course:

"The building of the Baltimore and Ohio Railroad into northwest Virginia was an event of great importance to that region, as well as to the State of Maryland."

* * * * *

"The Baltimore and Ohio Railroad was important for other reasons. It was the second railroad in America to use steam engines and the first to carry passengers and freight. Already there were a few railroads in England, but, for actual trade and travel, railroads were untried anywhere. It had not even been decided as yet whether stationary steam engines, horses, or locomotives would prove to be the best motive power."

"It was a great day for Maryland and northwest Virginia when, on July 4, 1828, Charles Carroll of Carrollton laid the first stones for the Baltimore and Ohio Railroad. Carroll was the last surviving signer of the Declaration of Independence. In speaking of his services in beginning the railroad, Carroll said, 'I consider this among the most important acts of my life, second only to my signing the Declaration of Independence, if even it be second to that.' At the same time, not far away, John Quincy Adams, President of the United States, was lifting the first shovel of dirt from the proposed Chesapeake and Ohio Canal. Despite the adverse reports of engineers, the canal was still regarded by Marylanders and Virginians alike as the most promising solution of their transportation problems."

"It was more than twelve years after that memorable Fourth of July before the railroad reached Cumberland, Maryland. Meanwhile, branch lines had been built to Washington, D.C. and to Frederick, Maryland. During this times the Baltimore and Ohio Company had solved the problem of a suitable motive power. Horses were used to first draw passengers and freight in box-like cars. Then Peter Cooper, an inventor living in the State of New York, called attention to the steam engine, which was already coming into use. Although the railroad directors were not greatly impressed by Cooper's suggestion, they permitted him to experiment. In 1829, he built a locomotive, called the *'Tom Thumb,'* which was then put to use the following year between Baltimore and Mount Clare, Maryland."

"Although the *'Tom Thumb'* had lost in a race with a horse, the owners were impressed by the engine's performance and offered a prize for an improved engine."

* * * * *

"The *'Lafayette,'* built in 1837, was the first American-built locomotive to have a horizontal boiler, such as all locomotives have today. It is the use of this type of boiler which has made possible their large size."

* * * * *

"The line was completed to Cumberland on November 5, 1842. The Baltimore and Ohio Company planned at once to extend its track to St. Louis and to New Orleans."

* * * * *

"In its plan to reach the West at once the Baltimore and Ohio Company was disappointed. It was not until 1847 that the Virginia Assembly gave the railroad company permission to extend its lines to the Ohio River. The Assembly refused the railroad permission to extend its lines beyond Cumberland because it wished to make western Virginia, and for that matter the entire West, tributary to Richmond and Norfolk. Pittsburgh took advantage of this opportunity to advance her own interests and tried to bring the railroad to her gates."

* * * * *

"The building of the railroad was resumed in 1848 and it was completed to Fairmont four years later. This section of the railroad required the use of bridges spanning streams and tunnels piercing mountains. By an engineering feat remarkable for its time, the Alleghenies were conquered. The crest was crossed near Altamont, Maryland, at an elevation of 2,622 feet. Then the road descended across the glade country to the Youghiogheny River country and climbed thence along Briery Mountain to Cranberry Summitt, with an elevation of 2,549 feet. This point is now Terra Alta, West Virginia. The descent to Cheat River, in a south-westardly direction, was rapid."

"A more remarkable feat was the completion of the famous tunnel at Tunnelton in Preston County, (West) Virginia. At this point the railroad came against a stone wall. As there was no way around, the mountain had to be pierced. The tunnel was 4,100 feet in length and at the time of its completion was the longest in the world. It took almost three years to complete this work, but while it was in progress the railroad track to the westward was being advanced. For this purpose a temporary track with an ascent of five hundred feet to the mile was laid across the mountain."

* * * * *

"The first railroad train to reach the Ohio River over a continuous track arrived in Wheeling on January 1, 1853. Celebration of the event did not begin until January

Thomas Swann, President and Benjamin H. Latrobe, Chief Engineer of the B&O were most responsible for building the line westward across the mountains, completed on New Years Day in 1853.

10, following. In attendance from the East, coming by way of the railroad, was a distinguished company which included the governors of Maryland and Virginia and members of the legislatures of these states. The legislature of Maryland had adjourned for the occasion. Benjamin H. Latrobe was engineer of the road, and Thomas Swann was President."

* * * * *

"With the completion in 1857 of other railroads, the Baltimore and Ohio Company was able to make through connections from Baltimore to St. Louis, with changes only at Grafton, Parkersburg and Cincinnati."

* * * * *

"The 'coming of the B&O' caused villages and settlement sites to grow into towns and cities. It developed industry, mining, and lumbering, and doubled and even tripled land values in the interior. The most immediate effect, however, was upon traffic on the Ohio River. There, for the first time, daily packets entered the trade between Wheeling and Louisville. The boats used were 'the finest that ever plied the waters of the upper Ohio in the line arrangement.' Their average capacity was in excess of eight hundred tons."

* * * * *

Two members of the Coal Development Office were featured in pictures showing employees of the company in the annual report.

Tom Hayes, our special representative in the New River field, was shown talking to Ray Tolliver at his mine office in the Coal River Field. Jack Davis had replaced the retired Earl Kiser as our representative in the Kentucky and Logan Fields was shown reading the Funnies to his three children. It is sad to relate that Earl died of a ruptured brain embolism only four months after retirement. He had been a good and loyal friend to me and I knew I would miss him.

Guy Hoover suffered a heart attack and decided to retire as he did not relish the job of managing our office. Right away, I received a brief personal letter from Walter Tuohy putting me in charge with the title of Engineer of Coal Properties. The letter stated that I was young for the job, but I was to be given a chance. The fact that my three predecessors were elderly men obviously made me look very young at age 42.

One of my first assignments was to attend a meeting of all Coal Traffic agency heads in Cleveland. When we duly assembled, Greg walked into the room and announced that our management was seriously considering merger with the B&O and he wanted our suggestion for reasons that such a move would be good for C&O. Most of the older heads present were horrified at this idea and immediately started to enumerate their low opinion of the B&O. Greg soon stopped any further complaints by raising his hand and stating emphatically that only positive comments were acceptable. A secretary was called in to list every conceivable good thought that anybody could come up with.

I had written a note to Greg soon after the Virginian N&W merger suggesting that we consider taking over the Western Maryland Railroad which I knew was profitable with considerable coal traffic. Greg replied that it was not feasible because it was owned and controlled by the B&O. I often wondered if my suggestion had triggered the new development.

Shortly after I took over the office, a staff member of the Finance Department, John Feagler, was sent to our office. He explained that he was ordered to give me some pointers on executive expectations and behaviors. He turned out to be an interesting and helpful individual. I took him on a trip to see some of the coal fields and introduced him to several coal operators. I finally deduced from his comments and suggestions that I was being groomed for executive responsibilities.

In July of that year, several members of our department were dispatched to large cities to try to persuade B&O stockholders to agree to our offer of a stock exchange. I was assigned to go to Philadelphia with Speed Gray who was an Assistant to Greg DeVine in Cleveland. Speed was a former operating trainmaster at Newport News with an extroverted personality who because of his experience was promoted to the coal department to look after all aspects of the export coal movements and lake coal.

I met "Speed" at the Philadelphia airport and we proceeded to one of the largest hotels. For three days we followed up on a prepared book of B&O stockholders. Most of them were receptive except a local stockbroker who along with several clients were major stockholders. He would hardly listen to us and ushered us out of his office. "Speed" called Walter Tuohy on the phone that afternoon and reported our experience with this intransigent individual who was approached later by the boss himself. Our evenings consisted of several cocktails and a dinner at one of the best restaurants in town. One of them, Bookbinders, was a seafood restaurant where I experimented with a dinner of steamed clams which were too rich for a hillbilly novice.

In the list of officers, two new appointments appeared; Hays Watkins as Treasurer and Robert C. McGowan as Comptroller.

Early in the year, Ed Haessler was transferred to Cleveland where it was decided he could be more effective in redefining our sidetrack agreements. His secretary, Frances Huddleston was added to our local force.

As President of the local chapter of the West Virginia Society of Professional Engineers, I decided to ask Raymond Salvati who lived in Huntington and was President of Island Creek Coal Company to be our speaker. I made a visit to his office and he accepted saying that the only reason he didn't belong to our group was that he was never asked. We had the largest turnout ever.

John F. Kennedy was elected President of the United States after spending considerable time in the coal fields of West Virginia campaigning.

His motive was to prove that a Protestant State would support a Catholic candidate. I ran into his entourage in the New River Field and got a lot of waves and smiles.

Cyrus Eaton had started holding annual international conferences at his farm in Pugwash, Nova Scotia. The idea evidently was to get east and west national civic leaders to sit down together in the interest of world peace. Willis Cook who was our PR man in Huntington at the time, accompanied the Chairman and lived in his office car and handled press relations during the conference.

On one of my visits with Joe Fish in Logan, he complained to me about the fact that the white numbers on some of our cars had worn off. Without this number, it was impossible to have the car traced. The numbers were painted with a special white paint that sluffed off as it oxidized which made the number stand out more but all the paint eventually wore off. The paint was called "self-cleaning" by the department. I wrote a note to Greg suggesting the use of bright yellow paint since it was one of our company colors. Several months later coal cars were sporting yellow numbers and they still do today.

Several changes were made in our office procedures to make its operation more efficient. A staff meeting was held every Monday morning at which time general operations were discussed and each staff member discussed his travel plans for the week. A list was then prepared to show where everyone on the road could be contacted. In addition to the field representatives, each younger engineer was assigned a territory for which he was responsible. In the manner of Dr. Lawall, I maintained an open door policy. The filing system was greatly improved by being completely re-vamped.

The railway officers were constantly being bombarded by the coal producers for lower rates. The energy market has always been extremely competitive. In the late fifties, Consolidation Coal Company used the threat of coal pipelines in an effort to reduce rates. They finally built a pipeline from their big surface mine at Cadiz, Ohio, to a utility plant in Cleveland. In one of those unforeseen oddities of life, my own brother-in-law, William Halvorsen, was the General Manager of the project. Greg and I went to see the installation and could see with our own eyes that powdered coal was being delivered as a slurry on the shore of Lake Erie. As soon as the pipeline proved feasible, Consol had a bill introduced in the West Virginia legislature to authorize the construction of a coal pipeline from one of their mines in West Virginia to a coal pier near Norfolk, Virginia. The C&O really took a dim view of this proposal which would detract substantially from our most profitable coal delivery route. Our company started to organize all possible opposition to the idea and encouraged its employees to fight against the effort. Finally, the legislature called a public hearing in the House of

Delegates' Chamber in the State Capitol. It was held in the evening to accommodate all working people. The C&O had one of our lawyers in the law firm in Huntington which represented us to make our case. Amos Bolen was a very impressive speaker and captivated the audience with his orating. I was sitting near the speaker's platform with several other Company Officers and Amos grabbed a bucket from under the podium. Holding it high above his head, he shouted, "Even if you carried a bucket of coal slurry to the east coast, it couldn't be sold without the permission of Consolidation Coal Company."

There were many more coal miners in West Virginia than railroad workers, so the legislature passed the bill to permit the construction of a coal pipeline. Fortunately for the railroads, the State of Virginia didn't want any pipeline crossing their state and stopped the project.

The railroads finally decided to reduce their rate for hauling coal from Cadiz to the lake and Consol moth-balled their pipeline.

DIRECTORS

Cyrus S. Eaton . . . . Cleveland, Ohio
Chairman of the Board

Walter J. Tuohy . . . . Cleveland, Ohio
President

Herbert Fitzpatrick . . . . Huntington, West Virginia
Attorney, former C&O Chairman

Barnum L. Colton . . . . Washington, D.C.
President, The National Bank of Washington

Robert J. Bulkley . . . . Cleveland, Ohio
Attorney, former US Senator

Cyrus S. Eaton, Jr. . . . . Cleveland, Ohio
Mining Development, Investments

Roger H. Ferger . . . . Cincinnati, Ohio
President, Publisher, Editor, The Cincinnati Enquirer

Dr. Fay A. LeFevre . . . . Cleveland, Ohio
Chairman of the Board of Governors, Cleveland Clinic

M. S. Fotheringham . . . . Steep Rock Lake, Ontario
President and General Manager, Steep Rock iron Mines Limited

Erwin H. Will . . . . Richmond, Virginia
Chairman of the Board and Chief Executive Officer,
Virginia Electric and Power Company

OFFICERS

John E. Kusik . . . . *Senior Vice-President-Finance*

M. Irvin Dunn . . . . *Senior Vice President-Operations*

Gregory S. DeVine . . . . . *Senior Vice President-Coal Traffic & Developement*

James E. Doyle . . . . *Vice President-Merchandise Traffic*

Joseph C. Kauffman . . . . *Vice President-Law*

Owen Clarke . . . . . . *Vice President-Labor Relations, Purchases and Stores*

Hays T. Watkins . . . . *Treasurer*

Robert C. McGowan . . . . *Comptroller*

T. Howard Keelor . . . . *Secretary*

☆

1961 - B&O MERGER HEARINGS

As had become the custom, the annual report opened with the letter to stockholders from the Chairman and President which was released as the Flash Annual Report on January 2, 1962. C&O is unique in being able to report so early:

	1961	1962
Operating Revenues (millions)		
Coal and Coke	$144	$157
Merchandise	149	162
Other	23	25
Total Operating Revenues	316	344
Expenses, Taxes, etc.-Net	286	302
Working Capital at Year End	50	54
Net Income	30	42
Earned per Common Share	3.60	5.15
Dividend Paid per Common Share	4.00	4.00

The letter elaborated as follows:

"In the last few months of the year just closed, the nation's economy was moving encouragingly upward. C&O activity advanced in step, with total earnings in the last five months substantially greater than in the first seven months of the year."

* * * * *

"Baltimore and Ohio-Chesapeake and Ohio affiliation progress was the highlight of the year. We continue to be enthusiastic about the benefits which this partnership would bring to Chesapeake and Ohio shareowners, through opportunity for growth in their investment. At the same time, this joining of C&O strength and B&O potential would advance the interests of B&O shareowners and the employees of both companies, and bring the needed strength to an important segment of the railroad industry in the East. It is our hope that Interstate Commerce Commission approval of B&O-C&O affiliation will be given in the first half of this year."

* * * * *

"Merchandise Traffic 1961: Revenues were up strongly as the year ended. Piggyback, the carrying of trailers on flat cars, was an especially bright spot, with volume 14 percent greater than in 1960. Transport of new automobiles in bi-level and tri-level railroad cars also advanced, with nearly 11,000 carloads moved."

"Indicating the continuing desirability of C&O's territory and service, 83 industries-19 more than in 1960-picked sites for new or enlarged plants on Chessie's lines. From them will come an anticipated $2.3 million in annual revenues."

* * * * *

"Coal 1961: Movement of bituminous coal to users in the United States and Canada held up well, although the delay in the economy's pick-up was felt. With industry at lower levels, demand for coal as a raw material and source of power declined. Thus, national coal production of 400 million tons was down from 1960's 416 million tons. Overseas shipments for export were less than in 1960."

"Plans to build a coal-carrying pipeline were abandoned by a large Michigan power company after the comparative costs of railroad and pipeline haulage of coal were studied and the railroads arranged a long-term incentive rate."

* * * * *

"Progress 1961: C&O grew in strength and ability to serve. An electronics communications system. faster and more efficient, was placed in service, and more miles of line were placed under the electronics supervision of Centralized Traffic Control. At Newport News, Virginia, work was begun on Coal Pier 14 to make it capable of loading more speedily than ever the super-colliers now transporting coal on the high seas. At Toledo, C&O began expanding its huge Great Lakes coal trans-shipping terminal."

"Outlook: Improvement of C&O strength and competitive ability will continue. High point of the year will be installation of the most advanced electronic data processing center in the railroad industry."

* * * * *

The report continued with a long letter about the coal traffic segment by Greg DeVine:

"The year ahead shows every prospect of being a better one for the coal industry. And a better year for coal is always a better year for C&O."

* * * * *

"In 1961, 400 million tons of coal were mined in this country, down from 416 million tons in 1960. C&O felt the effect. Coal loaded at mines on our lines constituted nearly 11 percent of national production, compared to almost 12 percent in 1960. Total movements over C&O from all sources were 920,000 carloads against one million carloads the year before."

"For the most part, C&O's shipments to users in the United States and Canada held up well. Although export business generally slipped under the 1960 mark, there were excellent signs of improvement in some major overseas markets."

"Japan, especially, has become a bigger customer for American coals. Japanese industries bought over 6 million tons in 1961 and plan to buy more in 1962."

* * * * *

"C&O's car fleet, already one of the largest in the industry, is being expanded and modernized. C&O shops have begun construction of 2,000 new cars of 85-ton capacity and 500 70-ton cars and a renovation program for 4,500 existing cars."

* * * * *

"In 1962, C&O data processing specialists will put new electronic computers to work for the Coal Traffic Department. From this new fact-finding service, we will be able to supply information to customers faster than ever before and further improve our department's internal operations."

"We are redesigning Pier 14 at Newport News, our Atlantic Ocean terminal for the export trade. When the work is completed in 1962, Pier 14 will be able to load as much as 9,000 tons of coal an hour-a 50 percent increase in its capacity-and handle the largest coal trade, as well as the even larger vessels planned for the future."

* * * * *

"One problem for everyone in the coal, oil and natural gas and railroad industries is the continuing imports of foreign residual oil into this country, particularly on the eastern seaboard. This waste product of the petroleum refining process can be dumped anywhere at any price under the cost of domestic fuel. We, along with others concerned with the problem, are urging the federal government to increase restrictions on these imports."

* * * * *

A page in the report outlined the merger news:

"On January 15, 1962, Chesapeake and Ohio filed with the Interstate Commerce Commission a brief summarizing its case in support of its application to control B&O as a step to merger,"

"C&O concluded its brief with this statement: 'With full recognition of the Commission's need thoroughly to consider this important matter, we urge that every reasonable effort be made to reach a decision as quickly as possible.' Among the points made in the brief were these:"

"... Blending of financially strong C&O with B&O os highly in the public interest, meets a basic purpose of the Interstate Commerce Act and is bound to strengthen the railroad picture in the East."

" . . . Substantial benefits to C&O will result from the affiliation through traffic diversification, major savings and an opportunity for C&O to employ its financial resources in highly-productive, income-producing projects."

"... Affiliation will redound greatly to the advantage of labor. C&O's program for rehabilitation and modernization of B&O will create jobs."

* * * * *

"On the same day that C&O filed its brief, the New York Central Railroad announced that it was withdrawing its opposition to the B&O-C&O affiliation, and petitioned the ICC for formal permission to do so. A few days earlier, Central and the Pennsylvania Railroad had announced that they had reached an agreement to seek a merger of their two properties."

* * * * *

The report concluded with some references to passenger service:

"Faster mail service via C&O passenger trains was the result in October, 1961, of matching C&O's Roadrailer with Detroit's new automated post office."

* * * * *

"Twenty Roadrailers, formerly called Railvans, now operate in Michigan."

"During the year, C&O's streamlined trains and Lake Michigan trainferries carried 844,000 passengers, over 18,000 of them traveling to and from the Greenbrier. Passenger service revenues of $13.7 million did not meet costs."

* * * * *

Mel Triolo replaced J. W. Coley as Secretary-Treasurer of the Logan Coal Operator Association after a long career as Union Organizer and later as Safety Director of Jewell Ridge Coal Company. I often met with him whenever I was in the Logan field.

In June, I was advised by Greg DeVine to come to Washington where our forces were housed in the Statler-Hilton Hotel to participate in the hearings. My responsibility was to refute any claim by the New York Central and Pennsylvania Railroads that our merger would dominate the availability of coal reserve in the Eastern United States. I knew right away that we had an ace that would be most helpful in the battle.

My father, Glenn R. Toothman, was working for the Bureau of Mines in Pittsburgh. He had advised me sometime before that there was a government sponsored coal reserve study of the entire United States in the Bureau's library and was available to the public. He knew that part of my job was to keep up to date records of coal reserves available to the C&O and an in-house study has been initiated before I came to the company. It was an invaluable tool in persuading coal operators to open new coal mines on our railroad. The study was performed and up-dated on a continuous basis by Gates Engineering Company in Beckley. This arrangement helped us in the fact that it was performed by an outside agency and therefore could not be discounted as self-serving propaganda.

I immediately put in motion the acquisition of the several volumes of information published by the Federal Government with the title of: *The Liquid Fuel Potential of the United States.* The study was made under the direction of the Corps of Engineers and they in turn contracted the Paul Weir Company, a nationally respected mining engineering firm in Chicago to perform the actual work. Close examination proved to us that the study was authentic because in the areas covered by our own reserve information, the results were strikingly similar. Having the information permitted us to use government backed statistics without revealing any of our own reserve information which was considered to be confidential and withheld from our competitors.

Jim Lilly and I worked diligently for the first week in Washington developing a study to show a comparison of our merged company reserves with those available to both New York Central and the Pennsylvania Railroads.

The opposing railroads had very early in the proceedings submitted testimony into the record of the hearings, by virtue of a reserve study prepared by a consulting engineer by the name of McCarthy from Pennsylvania. His study was based on reserve information taken from geological survey publications of the various states in which the eastern railroads operated. He contended that C&O-B&O would have access to twice as much coal reserves as all the other eastern railroads including the New York Central and the Pennsylvania which also controlled the N&W.

After we finished our study, it showed that C&O-B&O would have access to about the same amount of reserves as the other three combined. The morning after we gave this information to DeVine, I was advised at breakfast to report to the main entrance of the hotel to meet Walter Tuohy. When I got there, he invited me to accompany him in a cab to the ICC building where the hearings were being held. On the way he told me he appreciated our efforts on the reserve study, but he believed our case would have to be stronger because the examiners approval could turn on my testimony. Needless to say, Jim and I went back to the drawing board and modified a couple of our assumptions after reconsidering the whole matter very carefully. This reduced the reserves available to railroads because they were close to navigable rivers or trucked. This time we came up with less reserves than our opponents railroads much to the relief and satisfaction of our management.

There was a very astute lawyer in our company who worked with the coal department. His name was Hewitt Biaett and he had obviously become a close friend with Greg as evidenced by spending a lot of time together outside of the hearings. A young lawyer by the name of John Hanifin worked for Hewitt and he was assigned to write my testimony. We worked together for one whole day preparing my statement that I presented as the last witness for our side to conclude the C&O/B&O case. I felt very proud of my statement because I was confident it would clinch our case. My statement is excerpted as follows:

I.C.C.F.D. Nos. 21160-21161
Prepared Statement
Witness: Fred R. Toothman

Q. Have you analyzed the testimony and exhibits of Witness McCarthy that have been introduced in this case?

A. Yes, I have.

Q. Have you during the course of your normal duties with the C&O undertaken a study of the coal reserves in the eastern and mid-western coal fields?

A Yes, I have. This is a routine function of my office and I have studied the coal reserves in the geographical area covered by Witness McCarthy's testimony and Exhibits No.s. H-287 - 290.

Q. Have you analyzed the testimony and exhibits of Witness McCarthy dealing with the coal reserves tributary to the railroads in the areas they serve?

A. Yes, I have, an in my opinion Witness McCarthy grossly overstated the coal reserves that are tributary to the railroads he studied, particularly the C&O and the B&O. This overstatement occurs because he included not only measured and indicated resources, but also inferred reserves as being recoverable coal. Moreover, he did not utilize the best available source of information for his study.

Q. Will you explain what you mean?

A. The initial source of information for any study of coal reserves is to be found in the individual state geological surveys which have been made with the assistance of the Federal Government. Those surveys are geologists' appraisals of coal reserves and from the standpoint of a mining engineer must be classified as raw data. Although the source materials used by Mr. McCarthy have important geological value, he did not rely on the only study that has been compiled by nationally recognized mining engineers which is the outstanding source of information for any realistic study of recoverable coal reserves.

Q. What study did you refer to Mr. Toothman?

A. I am referring to *"The Synthetic Liquid Fuel Potential of the United States,"* a report prepared for the Bureau of Mines, Department of Interior, by Ford, Bacon & Davis, Engineers, in 1952. This study contains a complete and detailed evaluation, that is adapted for a mining engineer's use, of certain of the natural fuel resources of the United States, particularly coal reserves. This study took the information that had been published by the state geological surveys as a point of departure and assessed the reserves in terms of those that are recoverable by presently known practical methods of mining.

Q. Will you state in detail the specific areas of error you find in Mr. McCarthy's testimony?

A. The use of only the state geological surveys overstates the economically recoverable coal reserves because the measured, indicated and inferred reserves are included, whereas the Ford, Bacon & Davis study shows only the reserves that can be extracted by practical mining techniques. Coal beds less than 24 inches in thickness were excluded in the Ford, Bacon & Davis study as not being recoverable. Further, the Ford, Bacon & Davis study recognized that information concerning inferred reserves alone is not sufficiently substantiated to support an investment in a modern mining operation.

Q. Will you indicate any other way that the McCarthy study overstates the coal reserves that are available for railroad transportation?

A. Mr. McCarthy, according to his testimony, made the basic assumption that all coal reserves, measured, indicated, and inferred, are tributary to or available for railroad transportation. The Mineral Industries Survey, published by the United States Department of Interior, shows that the percentage of coal carried by railroad compared to the total amount of coal produced in the United States has declined from 81 percent in 1950, to 73 percent in 1959. In other words, 27 percent of all coal produced in 1959 was not transported by a railroad.

Accordingly, I consider the study of coal reserves made by Mr. McCarthy to be unrealistic, for it not only relies on inferred reserves but it assumes the premise that all coal reserves existing in the ground are available to railroad transportation. This is not the case.

Q. Mr. Toothman, is there any other reason why you consider the coal reserves Witness McCarthy has allocated to the NYC and other railroads to be inaccurate?

A. Yes, there is. As Mr. McCarthy has stated in his testimony, the study of coal reserves he made were of fields located in Ohio, Pennsylvania, Maryland, West Virginia, Virginia, and eastern Kentucky, tributary to a selected group of railroads. His failure to include the coal reserves tributary to these railroads that operate in Illinois and Indiana does not give a complete picture of reserves, particularly those of the NYC, and results in subsequent distortion of the percentage allocation he makes to the several railroads included in his study.

Q. Mr. Toothman, have you made a study of all the recoverable coal reserves available to the same railroads that were listed by Witness McCarthy, based on a mining engineer's appraisal of the potential railroad traffic?

A. Yes, I have, but since my analysis embraces all of the recoverable coal reserves that are available to the selected railroads studied by Mr. McCarthy, I have included those in Indiana and Illinois.

Q. Will you please state the method and procedures you employed in making this survey?

A. For reasons I have already explained, my study is based on the Ford, Bacon & Davis report. The results of my analysis are shown on a statement entitled *Recoverable Coal Reserves Available to Certain Railroads and Others in Selected States.*

It should be noted that although the C&O and B&O have available 10,643,000,000 tons of recoverable coal reserves, the remaining 72 percent, or 26, 911,000,000 tons are available to other railroads, other forms of transportation, or captive industries.

Q. Have you undertaken any further measurement of coal reserves tributary to the C&O, B&O, and NYC?

A. Yes, I have. I have determined that, by using the 1960 originated tonnage of each of the three carriers, and the total recoverable reserves shown on my statement, all three railroads - C&O, B&O, and NYC - have reserves of recoverable coal that will last in excess of 100 years.

I.C.C.F.D. Nos. 21160-21161
Exhibit No. ____________
Witness Fred R. Toothman
(Iden. No. FRT-1)

Recoverable Coal Reserves Available to Certain Railroads and Others in Selected States as of January 1, 1961
(Million of Net Tons)

	STATES								
Available to	Ohio	Penn.	W. Va.	Va.	East Ky.	Maryland	Indiana & Illinois	GRAND TOTAL	% OF TOTAL
C&O	249	-	2,972	10	2,099	-	-	5,330	14
B&O	399	1,406	2,959	-	-	-	549	5,313	14
W. Md.	-	-	312	-	-	100	-	412	1
Penn.	531	3,584	168	-	-	-	715	4,998	13
N&W	10	-	2,731	1,552	721	-	-	5,014	13
DT&I	177	-	-	-	-	-	-	177	1
Nickel Plate	197	-	-	-	-	-	600	797	2
NYC & P&LE	293	1,267	417	-	-	-	902	2,879	8
Others*	1,230	500	159	625	971	41	9,108	12,634	34
Total	3.086	6,757	9,718	2,187	3,791	141	11,874	37,554	100

* Includes barge, truck, pipeline and conveyor transportation, mine-mouth power plants, other industries, and other railroads.

When I started making regular trips to Cleveland to report to my boss, the company was operating a overnight sleeping car from Huntington to Cleveland and return daily. The C&O passenger train took it to Fostoria, Ohio, where it was picked up by the Nickel Plate and delivered to Cleveland. The car had bedrooms for two people in the middle and a single compartment on each end. Since there were very few passengers on the car, I usually got to use a bedroom and many times the porter and I were the only ones on the car. I never slept very well on the first night but was so tired on the return that I usually slept very well.

After DeVine came aboard, we were advised to fly. For me this required an Allegheny flight to Pittsburgh where I transferred to Northwestern for the second leg to Cleveland. Northwestern had bigger planes and much better service and I always rode in the first class section. The stewardess would serve a cocktail even before the flight left the gate.

☆

1962 - MERGER APPROVED AND EARNINGS IMPROVE

The annual letter was brief and was preceded by a welcome to former B&O shareholders:

"A special greeting to the new C&O shareowners who were B&O shareowners until February 4, 1963, accompanies this Flash Annual Report, giving the basic results for 1962. It was sent to C&O shareowners on January 2, the first working day of 1963."

* * * * *

The letter started out with the merger news:

"The good news C&O has awaited-approval of its affiliation with the Baltimore and Ohio Railroad-came on the last day of the old year."

"The Interstate Commerce Commission just two days ago approved as consistent with the public interest C&O's control of the Baltimore and Ohio, the nation's first railroad and one of its most important."

* * * * *

"Higher earnings of $35.4 million reflected a number of favorable factors: increased activities in industries which use our coal and generate merchandise movements, benefits from new tax laws and rulings and continued control of operating and administrative expenses."

"Improvement in coal traffic was notable, revenues of $157.7 million exceeding 1961's $143.7 million by 9.7 percent. Shipments of 8.8 million tons for overseas export through our Newport News piers were up 29 percent. C&O's originations-the coal loaded at mines along our railroad-amounted to 11.4 percent of the national production of 419 million tons, up from 10.6 percent in 1961."

"Merchandise freight traffic revenues of $156.9 million were 5.4 percent better than 1961's revenues of $148.9 million. From the standpoint of growth, the highlight was piggyback-carrying truck trailers on flat cars-and new automobiles on special double-deck rack cars. Nearly 60 percent more piggyback business was handled, bringing revenues of over $3 million. Another $3 million was earned by the movement of new automobiles."

* * * * *

"C&O ordered 35 more diesel-powered locomotives, 2,000 coal hopper cars of 85-ton capacity and 325 specially-equipped cars for damage-free merchandise transport. Capital expenditures in the year totaled $50 million."

"Finances at year-end were excellent. Cash and short-term investments were $76 million, second highest in company history. Working capital was a record $55 million."

* * * * *

"If the Congress can now add enactment of a Transportation Act of 1963, based on the President's program, and if prompt approval can be given to railroad mergers and affiliations in the public interest, the transportation industry will at last be well on its way to a position of strength adequate to this nation's needs."

* * * * *

Highlights	1962	1961
Operating Revenues (millions)		
Coal and Coke	$158	$144
Merchandise	157	149
Other	24	23
Total Operating Revenues	339	316
Expenses, Taxes, etc.-Net	304	286
Working Capital at Year End	55	50
Cash and Short-Term Investments	76	63
Net Income	35	30
Earned per Common Share	4.30	3.60
Dividend Paid per Common Share	4.00	4.00

The rest of the annual report followed:

"At 12:01 a.m. on Monday, February 4, 1963, C&O formally took control of the B&O, as a step to merger. The affiliation was accomplished in a midnight ceremony in Baltimore in the portrait-lined, paneled board room of the nation's first railroad, founded in 1827. The ceremony was simple: a quiet signing of papers by officers of both companies and a transfer of stock certificates."

* * * * *

"The offer was to exchange one C&O common share for each one and three-quarter shares of B&O common and one C&O share for each share of B&O preferred. When the Exchange Offer closed on February 2, 1961, over 60 percent of B&O's outstanding shares had been assented."

* * * * *

"On May 1, 1962, Examiner John L. Bradford recommended that the C&O control of B&O be approved as in the public interest. On December 31, the ICC confirmed it."

* * * * *

"Between them B&O and C&O have total assets of $2.3 billion and had operating revenues in 1962 of $705 million. Together they are about the size of the New York Central, which has 10,218 miles of track and had 1962 revenues of $632 million."

* * * * *

"Already underway is a quarter-billion dollar improvement program for B&O. Cars and locomotives will be acquired, tunnels enlarged to pass piggyback cars and other high and wide shipments, piers and docks will be improved and new terminals will be built."

* * * * *

"B&O began its first railroad operations at Baltimore in 1830. C&O's original rail passenger, the Louisa Railroad, ran its first trains from Richmond, Virginia, six years later."

* * * * *

"In 1962, C&O moved a total of 62 million tons of coal, ten percent over 1961's 56 million tons. They produced revenues of $157.7 million, almost ten percent better then 1961's revenues of $143.7 million. And the increase were reflected in every facet of C&O's coal business-carloadings, originations, Great Lake movements and overseas shipments."

"C&O's good year, of course, depended first of all on the mines on its lines that produce the bulk of its coal traffic. In 1962, because they further strengthened their competitive position through improved operations and marketing, they increased their production by 12 percent, to 48 million tons from 43 million tons in 1961. The rise is doubly significant when compared with the less than five percent increase in the national production--420 million tons in 1962 verses 402 million tons in 1961.

"The greater strength of the C&O-served mines was particularly felt in C&O's export business, in which the new coal car fleet and the redesign of Pier 14 played such prominent parts. Foreign markets brought more American coal, and more of it from mines on C&O. By year's end, C&O had shipped through Newport News 8.8 million tons of coal, 29 percent more than the 6.8 million tons the port handled in 1961. They contributed $35.7 million to C&O's total coal revenues, compared to $28 million the year before."

* * * * *

A sad note was included about a former member of the Board:

"Herbert J. Fitzpatrick, of Huntington, West Virginia, a member of the Board whose associations with C&O went back sixty-seven years, died July 6, 1962, at the age of ninety. 'Mr. Fitz,' as he was affectionately known, was named Vice President and General Counsel of C&O in 1923 after having served the railroad as local counsel and then state counsel. In 1928, he became Vice President-Law of C&O, the Pere Marquette and Nickel Plate Road. He was elected to C&O's Board of Directors in 1930 and served as Chairman of the Board from 1937 to 1940."

"Two large power plants were completed on C&O which hold the promise of substantial new coal traffic for many years in the future. The Big Sandy Plant of Kentucky Power Company was completed and located on Louisa Fork mainline about twenty miles south of Ashland, Kentucky."

"Consumer Power's new plant commenced operations at West Olive, Michigan."

* * * * *

By this time, a Better Service Conference had been active in Huntington for several years. It had been started in the late fifties by an individual by the name of John Fuhrmann of Cincinnati, a public relations man hired by the company to serve several of these groups in different cities. The purpose of the Conference was to indoctrinate a group of middle management people in the need for deregulations of the railroads and to present the best possible effort at public relations in the local environment. J. B. Herring and I became charter members of the group. We were given a brief course in speech making and were prepared to make talks before local organizations. I had participated wholeheartedly in the effort and was the first Chairman of a model railroad show during the Christmas Season where the admission was a used toy or a small donation. The local Marine Reserves collected the toys, repaired them and distributed them to needy families. Prizes were awarded to exhibitions in two classes, junior and senior. These shows were very successful and are still being held in Huntington under different sponsorship.

My trips to Cleveland averaged about bi-monthly when Greg asked me to represent the coal department on a new inter-departmental committee chaired by John Marshall Jr., the son of the man who had been much revered for his railroad knowledge. The purpose of this committee was to explore better communications between departments and make suggestions to management for improvement. All other departments were represented. The Personnel Department representative was a young psychologist by the name of Norman Halperin who dominated these meetings. He was slight of build with a toothy smile and straight black hair. It was very obvious that he was extremely intelligent and was well aware of the fact. He was always very affable with me and I knew he knew all about me from my Western Reserve evaluation in the files.

This new responsibility meant that I had to be in Cleveland every month for a meeting. This extra travel in addition to my travel in the coal fields became very oppressive and I began to look for a way to get out of the assignment.

I was beginning to worry about my relationship with DeVine. I became aware of the fact that when I visited him to make my periodic reports that he never offered to shake hands and I always had to initiate the procedure. This really bothered me until a later event confirmed my suspicions once and for all.

The year was marked by another retirement in the person of my faithful secretary, Russell Blue. His departure left mostly female clerks in our office and the senior one was Fran Huddleston. I was aware that she was a good worker and she was most helpful to me all through the rest of my career. As Chief Clerk, Kenny Lemons was the only male.

Fred Palmer was a very successful contractor who lived in Huntington and did a lot of work for the railroad. He and I got to be good friends as fellow members of the local Society of Professional Engineers Chapter. He was a graduate of VPI and we traded barbs in the fall when our teams met on the gridiron.

The C&O for many years owned a large interest in the New River Company which was the largest coal company in the New River coal field at one time. I started working with J. A. "Alex" Hunt on many matters in the area where our railroad had joint service with the Virginian Railway at many times. Alex had been president of the company at one time, but now acted as Board member and Senior Advisor with an office in downtown Beckley. He had a special liking for coal field maps and I frequently brought him a new one to study and refer to during our meetings.

All during my career, I tried to keep abreast of activities at my alma mater and participated in all the alumni activity to some extent. Joseph W. Leonard was the head of a research bureau in the School of Mines at WVU and made some great strides in coming up with innovative ideas for the coal industry. He later became the Dean of the College of Minerals and a long-time friend.

1963 - EARNINGS PROVE MERGER EFFECTIVE

The usual letter heading up the annual report first reported the merger news and other news followed:

"Earnings were greater. The figures in the highlights table tell the good story. Shareowners received the $4 dividend."

* * * * *

"The affiliation actually became effective on February 4, 1963, when C&O issued shares of its stock to B&O shareowners in exchange for B&O shares, of which C&O now holds almost two-thirds. Later in February, the C&O's Chairman, President and Senior Vice President for Finance became members of the B&O's Board of Directors. A number of men from both companies assumed joint responsibilities during the year, as cooperation and coordination planning intensified."

"Even before affiliation was formalized, the two railroads had started to achieve improvements in transportation services to the public. C&O assisted B&O making available freight cars, locomotives and other equipment totaling in value $100 million. In all, C&O's investment in B&O has reached $225 million."

"By its own vigorous efforts and with C&O cooperation, B&O turned a loss position in 1961 into a net-income position the next year. The commendable recovery continued through the year just ending, with earnings of $5.5 million. The outlook is for results even better in 1964."

"On December 16, 1963, the President of the C&O became also Chairman of the Board of the Baltimore and Ohio. In taking the Chairmanship of the nation's first railroad, he pointed out the appropriateness of the tie between two companies which trace their corporate lines to canal companies of which George Washington was founder and first President."

* * * * *

"Briefly, now, here is the story of C&O's traffic and operating progress in the last twelve months."

"Revenues were greater and, through efficiency and control of costs, net income was increased 17 percent. Gains were made in physical strength, financial vitality was kept at high levels and important steps were taken to insure future growth."

"Coal traffic advanced exceptionally, revenues rising by nine percent. Outstanding was the improvement in coal shipments for overseas. Through its huge ocean terminals at Newport News, Virginia, C&O shipped nearly eleven million tons of coal, 22 percent more than in 1962. This good level of export business is expected to continue."

"At home, too, the demand for coal was on the rise. Reflecting this, C&O's Great Lakes terminal near Toledo had a near-record year. More than 18 million tons of coal, carried to the terminal in rail cars, were loaded into vessels for trans-shipment to consumers in this country and Canada."

* * * * *

"Most dramatic was the persistent progress of piggyback and the movement of new automobiles on special multi-deck cars. The number and variety of piggyback services were expanded, with 16 percent more trailers carried and revenues increased by 38 percent. Revenues from carrying new autos went up 19 percent."

* * * * *

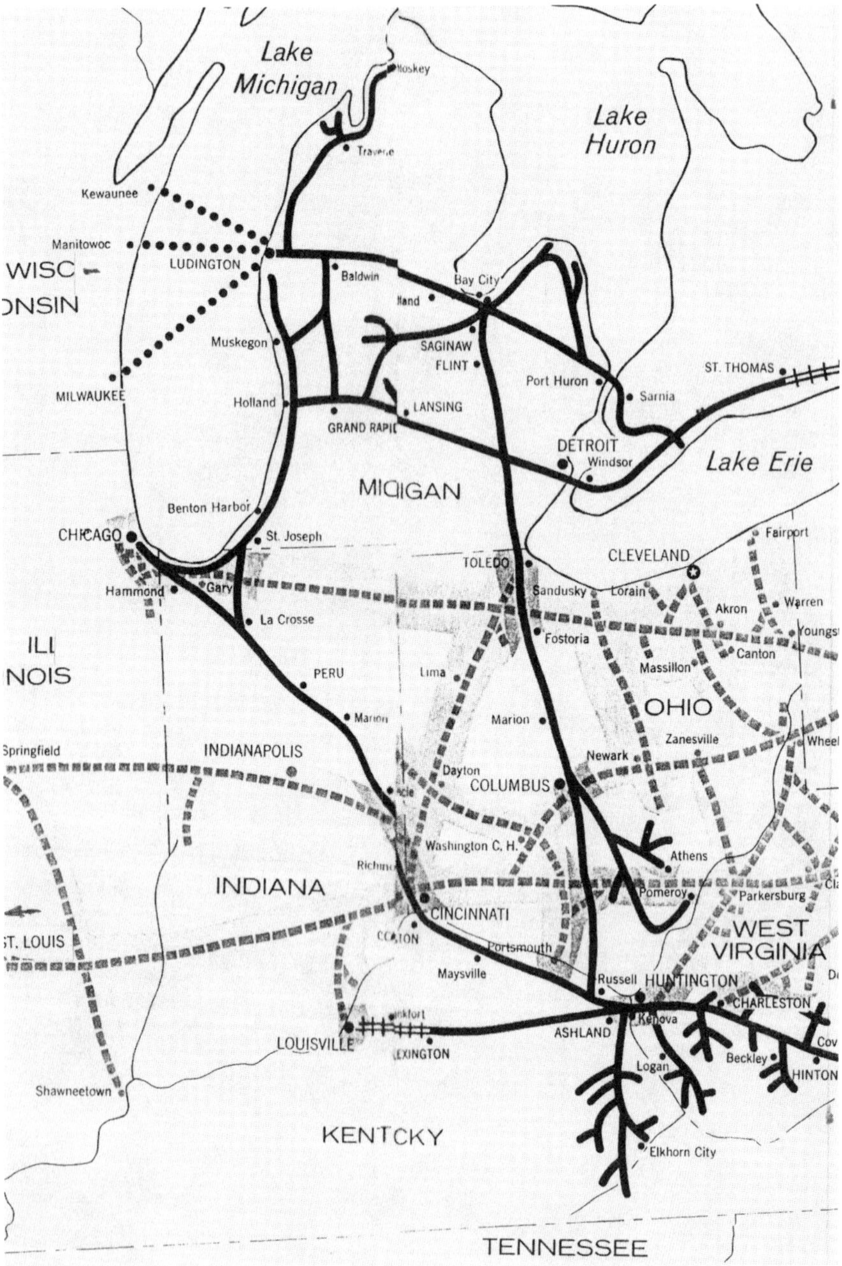
Lake Michigan
Lake Huron
Lake Erie
Kewaunee
Manitowoc
LUDINGTON
WISCONSIN
Baldwin
Bay City
SAGINAW
FLINT
Muskegon
MILWAUKEE
Holland
GRAND RAPI
LANSING
Port Huron
Sarnia
ST. THOMAS
DETROIT
Windsor
MICHIGAN
Benton Harbor
St. Joseph
CHICAGO
Hammond
Gary
La Crosse
TOLEDO
Sandusky
Lorain
CLEVELAND
Fairport
Akron
Warren
Fostoria
Canton
Massillon
ILLINOIS
PERU
Lima
Marion
Marion
OHIO
Zanesville
Springfield
INDIANAPOLIS
Newark
Dayton
COLUMBUS
Washington C. H.
Athens
INDIANA
Pomeroy
Parkersburg
CINCINNATI
WEST VIRGINIA
ST. LOUIS
Portsmouth
Maysville
Russell
HUNTINGTON
CHARLESTON
Kenova
ASHLAND
LOUISVILLE
LEXINGTON
Logan
Beckley
HINTON
Shawneetown
KENTCKY
Elkhorn City
TENNESSEE

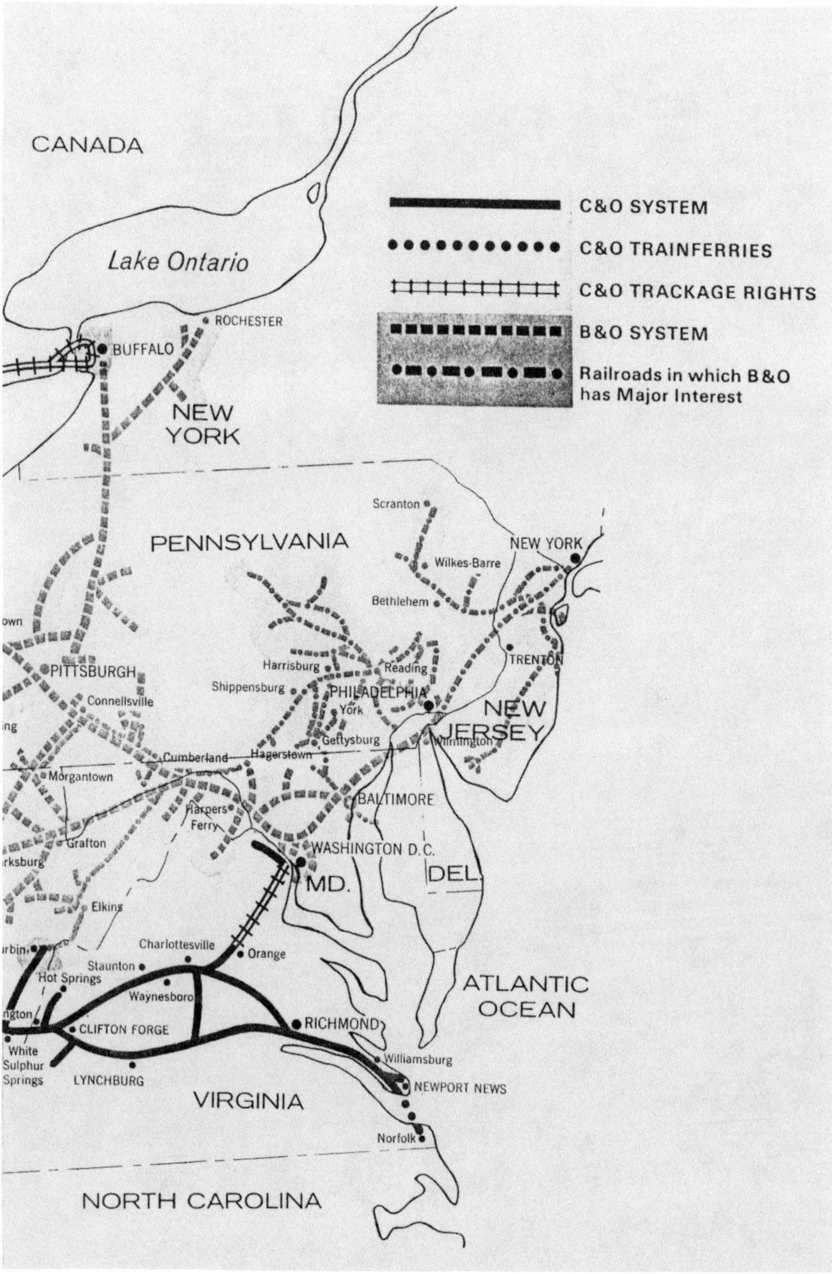
CANADA
Lake Ontario
C&O SYSTEM
C&O TRAINFERRIES
C&O TRACKAGE RIGHTS
B&O SYSTEM
Railroads in which B&O has Major Interest
ROCHESTER
BUFFALO
NEW YORK
Scranton
PENNSYLVANIA
NEW YORK
Wilkes-Barre
Bethlehem
PITTSBURGH
Harrisburg
Reading
TRENTON
Shippensburg
PHILADELPHIA
Connellsville
York
NEW JERSEY
Gettysburg
Wilmington
Cumberland
Hagerstown
Morgantown
BALTIMORE
Harpers Ferry
Grafton
WASHINGTON D.C.
DEL.
MD.
Elkins
Charlottesville
Orange
Staunton
Hot Springs
Waynesboro
ATLANTIC OCEAN
CLIFTON FORGE
RICHMOND
White Sulphur Springs
Williamsburg
LYNCHBURG
NEWPORT NEWS
VIRGINIA
Norfolk
NORTH CAROLINA

"Chessie added to its home-based sources of business in its richly diversified territory by helping industry locate 102 new plants there. The new facilities are expected to bring C&O $2.2 million in additional revenues. Since 1950, C&O has brought 1,240 new industries to its nine-state territory, from which it has gained an estimated $41 million in additional annual revenues."

Highlights	1963	1962
Operating Revenues (millions)		
Coal and Coke	$172	$158
Merchandise	164	157
Other	25	24
Total Operating Revenues	361	339
Expense, taxes, etc.-Net	320	304
Working Capital at Year End	41	55
Cash and Short-Term Investment	70	76
Net Income	41	35
Earned per Common Share	4.60	4.30
Dividend Paid per Common Share	4.00	4.00

"Record expenditures for equipment totaling $145 million were set on C&O for 1963 and 1964. This will put on the rails an additional 4,500 coal cars, including 'Super 80' hopper cars which can carry more than 80 tons each, 2,000 plain box cars in which the national shortage is particularly acute, 500 specially equipped and 121 diesel locomotives."

"With 50,000 C&O coal cars in service, no mine on the railroad in recent years has lost a day's work because of a lack of cars to load."

* * * * *

"Financial condition remained excellent. At year's end, all the basic indicators bespoke strength. Working capital was $41 million, and cash and short-term investments totaled $70 million. The ratio of total debt to total capital was 40 percent."

* * * * *

Further elaboration of news was carried in several articles in the report which are abstracted here:

"B&O completed 1963 with earnings of $5.5 million. In 1962, it had a slight profit of $1.6 million and the year before a deficit exceeding $31 million. Thus, from its low point in just two years B&O has made a $37 million recovery."

* * * * *

"Starting in mid-1961, the low point in B&O's fortunes, Chesapeake and Ohio extended assistance by making freight cars and locomotives available. Equipment made available then and later included 14,334 hopper cars, 1,275 box cars, 664 special-purpose cars, 124 diesel locomotives and 389 units of roadway maintenance machinery. Through 1963 the value of this help total more than $100 million."

* * * * *

"Most dramatic of B&O's rehabilitation projects was the enlargement of eighteen tunnels along an eighty-mile stretch of main line in West Virginia."

* * * * *

"This $8.5 million tunnel improvement program, completed near the end of 1963, was part of a $12 million system-wide clearance project. It gives B&O the shortest high car rail route between the East and the Southwest."

* * * * *

"'We are convinced that the planning function must be performed by key personnel on an accelerated basis without interruption by daily tasks,' Walter Tuohy, President of C&O, and Jervis Langdon, Jr., President of B&O, explained in a joint announcement."

* * * * *

"Men selected from both railroads have been appointed joint B&O-C&O officers with the title of General Manager of Planning in these three areas. They have their headquarters in Baltimore, maintain a joint staff and report to a policy board composed of top management of the companies."

* * * * *

"Chesapeake and Ohio on February 5, 1964, received a check for $65 million in payment for its 503-acre tract of coal docks and port facilities at Presque Isle, near Toledo, Ohio. The Toledo-Lucas County Port Authority bought the facility, all of which it then leased back to C&O for forty years, except for twenty-three acres on which new piers will be built."

* * * * *

"The $145 million spent in 1963-64 for new freight cars and locomotives will be the biggest bill C&O has ever paid for equipment in any two-year period."

* * * * *

"And new equipment costs less to keep in good repair. In 1963, 4.7 percent of C&O's freight car fleet was out of service for repairs, compared to an average of 7.2 percent for all Class 1 railroads."

* * * * *

"Four new U25B diesel locomotives do the same work as five of older units."

"Pier 14 and its sister, Pier 15, had one of their better years in 1963 and they are getting ready for a good one again in 1964. The nearly 11 million tons they handled was a 22 percent gain over 1962 and a 57 percent increase over 1961."

* * * * *

"C&O unit trains, solid trainloads of coal running between mines and C&O lines and Chicago steel mills, were started in 1963. This new, faster kind of transportation lets C&O reduce the Chicago rate on movements of 7,200 tons of coal. The trains are cheaper to run because the cars are larger, with capacities over 80 tons, the trains run directly between mines and mills without switching and adhere to strict loading an unloading time schedules."

"No matter how they heat their homes, millions of Americans consume a tremendous amount of coal-about a ton a year a household, to be exact."

* * * * *

"The end of this growth is nowhere in sight. By 1975, authorities predict, the electric power industry will be using as much as 450 million tons of coal annually. This is almost equal to today's United States coal production."

* * * * *

"Chessie offers piggyback service in its own trailers, as well as those belonging to trucking companies or shippers. In 1963, it added 200 new highway trailers to its fleet, bringing total strength to 800 units."

* * * * *

"In the year, C&O also entered into a partnership with thirty-nine trucking companies. It extends the railroad's piggyback operations into thirty states."

* * * * *

"Chessie's newest patrons include some as large as an automobile parts warehouse in Michigan that will bring C&O over 2,000 carloads of business a year and over $400,000 in revenues. There were others as small as a soap plant in Ohio which will mean about $1,000 annually for the railroad."

* * * * *

"John E. Kusik, Cleveland, was Chairman of the Finance Committee. Mr. Kusik is also a Director and Vice President-Finance of the Baltimore and Ohio Railroad. Until December 31, 1963, he had been Senior Vice President for Finance of C&O."

* * * * *

"Gregory S. DeVine was elected Executive Vice President in 1963. He continues his responsibility for C&O's coal traffic and development department."

* * * * *

"Own Clarke, was Vice President for Labor Relations, Purchases and Stores. Before joining C&O in 1958, he was Chairman of the Interstate Commerce Commission, Defense Transport Administrator and Chairman of the Public Service Commission of the State of Washington. He was born in Idaho and is a graduate of the University of Washington."

* * * * *

"Howard Skidmore was elected Vice President of both C&O and B&O in 1963 with responsibilities mainly in public relations."

* * * * *

"Shareowners are cordially invited to attend the railroad's Annual Meeting on Friday, March 20, 1964, at C&O's resort hotel, The Greenbrier, at White Sulphur Springs, West Virginia. A Shareowner House Party will also be held. For the first time, shareowners are being offered a four-day weekend beginning Thursday, March 19, or a three-day weekend beginning the day of the meeting."

* * * * *

There was a political development during the summer in eastern Kentucky that required a diversion in the routine of the Coal Development Office. Several local officials in the area started to make speeches about advocating the canalization of the Big Sandy River and Levisa Fork as far south as Pikeville, Kentucky. The main objective for such a project was to reduce the cost of shipping coal to the Ohio River and thus give the Kentucky coal operators a cost advantage over other districts, particularly southern West Virginia.

The C&O and N&W railroads took a very dim view of the proposal due to the obvious diversion of their coal traffic which such a project would entail. After a lot of publicity and agitation, the Corps of Engineers called for a hearing in Paintsville, Kentucky, for the opportunity for all interested parties to present their position, pro or con. The railroads decided to let Stuart Saunders, the diminutive but dynamic President of the N&W, present their case against the proposal. After several prominent citizens and local coal operators spoke in favor of the project, Stuart took the podium and delivered the most vigorous and slashing attack on the idea that could be imagined.

I along with two of my staff attended the hearing as representatives of the C&O. We were very much encouraged by Stuart's address. He quoted facts and figures to make his case that the project could not be cost effective, operation of barges would be restricted to very small loads, weather would close it down in the winter and further demonstrated that the whole idea was foolish to even be considered.

Stuart did such a magnificent job that enthusiasm for the project soon diminished in the area but even so, the local congressman was able to influence the Corps to make a preliminary study and report in another hearing in Pikeville in November. I also attended that meeting which had barely gotten underway when a message was delivered to the moderator of the meeting and he announced the assassination of President Kennedy and adjourned the proceedings. Evidently, the report that had been prepared by the Corps analyzing the project was unfavorable because the local support seemed to disappear and the proponents abandoned the concept.

The annual report contained a note that Walter Tuohy had been presented an honorary Doctorate from West Virginia University.

The year was the centennial for West Virginia which was made a State by President Lincoln in June, 1863. One of the celebrations was an opening football game for WVU in Morgantown against Navy. The game could have been arranged by the influence of Senator Byrd. Roger Staubach made a shambles of the Mountaineers by his passing and scrambling ability to win by a score of 60-0. I had attended with my parents as guests and my father remarked at the end of the game that at least we had the best band because Navy didn't send theirs.

Another retirement was conducted in the Coal Department when Roy Riedinger left. He had never seemed to get along with Greg and must have been ready to escape the corporate rat race.

I started attending some of the conventions of the Surface Mining Association of West Virginia and met Lawson W. Hamilton. He started punch and surface mining with his father and gradually became a big operator on the railroad. He was an outgoing personality and delights in being the "life of the party."

☆

1964 - WESTERN MARYLAND CORRALLED

The annual meeting was held at the Greenbrier on March 19, 1965 and a year of progress reported:

"C&O did well. Record revenues from carrying merchandise freight were earned. Revenues from transporting coal increased over the year before."

"Profits were up, too. Net income was $42 million, $1 million ahead of 1963."

"Shareowners received dividends of $1 a quarter, $4 annually, for the eighth consecutive year."

"The Baltimore and Ohio also had a better year. C&O's partner earned $7.2 million, an improvement of $1.7 million over 1963."

* * * * *

"A major management unification bringing together under joint officers most of the functions of the two companies, took place in December."

* * * * *

"An earlier step to management unification took place in October, when Walter J. Tuohy, C&O's Chief Executive Officer, became also President of the B&O. This followed the decision of Jervis Langdon, Jr., President of the B&O since 1961, to accept the position of Chairman and Chief Executive Officer of the Rock Island Lines." (Ed. Note: Langdon had been a top executive in C&O Traffic Department in the 1930's).

* * * * *

"Further unifying direction of the partner railroads, C&O President Gregory S. DeVine was elected in April a member of B&O's Board. And John E. Kusik, Chairman of the Finance Committee of C&O's Board, became at year end Senior Vice President of B&O, of which he had been a Vice President."

* * * * *

"On November 30, the B&O-C&O case for control of the Western Maryland as a step to merger opened before the Interstate Commerce Commission in Washington. The 844-mile Western Maryland operates in Maryland, West Virginia, and Pennsylvania, and is headquartered in Baltimore. B&O has held major, but non-voting, interest in it since 1927. Today, B&O and C&O own about two-thirds of Western Maryland stock."

* * * * *

'Rich bituminous coal fields in its territory provide the road with 45 percent of its traffic. Other business comes from steel, pulp and paper, cement, glass and chemicals, grain for export, and iron ore imports. In 1964, revenues of $47.3 million were up by $2.7 million over 1963 results. Net income was $5.3 million, $1.3 million above 1963."

* * * * *

"B&O acquired a 43 percent stock interest in Western Maryland in 1927 but has not been allowed by the Interstate Commerce Commission to vote the shares. C&O and B&O together now own about two-thirds of Western Maryland's stock, and have asked the ICC for authority to control the railroad as a step to merger."

* * * * *

"Returning to how C&O did in 1964, here is a closer look at the results."

"Earnings were equal to $4.72 a share, compared with $4.60 in 1963. Eleven percent of total revenues was carried through to net income."

"Total revenues of $373 million were up three percent over 1963. Expenses, totaling $331 million, were also higher by three percent, due mainly to the larger volume of traffic handled and greater employment costs."

* * * * *

Financial Highlights	C&O		B&O	
	1964	1963	1964	1963
Operating Revenues (millions)				
Coal and Coke	$174	$172	$103	$101
Merchandise	173	164	242	233
Other	26	25	37	38
Total Operating Revenues	373	361	382	372
Expenses, Taxes, etc.-Net	331	320	375	366
Working Capital at Year End	40	41	(8.9)	(9)
Cash and Short-Term Investments	67	70	29	22
Net Income	42	41	7.2	5.5
Earned per Common Share	4.72	4.60	1.89	1.21
Dividend Paid per Common Share	4.00	4.00	--	--

"Chessie's coal trains transported 53.7 million tons from mines on C&O lines, 1.2 million tons more than in 1963. Total revenues of $174 million were $2 million higher than the previous year."

* * * * *

"In 1964, the area showed vigorous growth, with C&O track-building forces engaged in eleven projects, building lines into new areas or extending lines to added mines. One of the track projects will ultimately add a million or more tons of coal a year to C&O's traffic."

"For equipment, C&O spent $78 million in 1964, the heaviest outlay on record."

"Acquired were fifty-six diesel locomotives and 7,700 cars for carrying merchandise and coal."

"An additional $12 million was spent for roadway improvement to extend electronic systems that direct train movements by remote control, to modernize the car-building shops and to erect a new rail-welding facility that will produce a smooth ride for people and goods moving over C&O and B&O."

* * * * *

"B&O's physical improvement continued with the assistance of C&O, which now has an investment valued at over $300 million in its partner. For equipment for B&O, including 102 diesel locomotives and 8, 100 freight cars, a record $88 million was spent. Another $12 million was expended for roadway maintenance and other capital projects."

* * * * *

"The Directors alternate their monthly meetings between Cleveland and Baltimore."

* * * * *

"1965 was only a week old when C&O's merchandise trains, which had made more money for C&O in 1964 than in any previous year, began making more exciting news. On January 8, C&O operated the first of a series of high-volume trains."

"Equipped with fifty-one giant covered hopper cars, the new trains carry 5,000 tons of wheat at a time in a high-speed shuttle service from Superior, Wisconsin, to the Pillsbury Company mills at Buffalo, New York. To give Pillsbury the best service and equipment available, C&O rushed into service thirty brand-new 100-ton hoppers designed for swift loading and unloading. From Superior, the trains roll over the Milwaukee Road to Chicago, where they are picked up by C&O for the 587-mile trip to Buffalo."

* * * * *

"C&O track forces are busily pushing lines into the West Virginia and Kentucky coal fields to tap new mines that will eventually produce about four million tons a year."

"These lines are signs of a healthy expansion in an area that already produces over 11 percent of the nation's coal. For instance, in Kentucky's Big Sandy field, tracks are being put down to two new mines that together will be producing 10,000 tons a day in full operation. Another Big Sandy line will eventually serve ten or more new mines."

* * * * *

"Record capital expenditures of $190 million by C&O and B&O in 1964 added new strength to both railroads."

"Most of the huge outlay-$166 million-was spent for new equipment, mainly freight cars and diesel locomotives. The remaining $24 million went for additions and improvements to tracks, structures, terminals, communications and signaling systems."

* * * * *

"Equipment acquisitions for both railroads together came to 158 locomotives and 15, 800 freight car of all types."

* * * * *

"The C&O-B&O Planning Department, established early in 1964 to concentrate in the operations areas of the two railroads, has been reorganized under a Vice President to extend planning emphasis to all C&O-B&O activities."

* * * * *

"Merchandise traffic, where so many of the newest wrinkles in railroad technology have come, saw more innovations in 1964. One of the biggest was the so-called 'high-cube car,' a super box car."

"The car is ninety-four feet long, longer than a bowling alley. It is seventeen feet high and has twice the cubic capacity of a standard fifty-foot box car."

* * * * *

"Another innovation is the hooded coil steel car. This special-duty car is designed to transport coils of steel, which despite their great size and weight, bruise easily and rust quickly.

* * * * *

"In 1964 B&O's freight revenues climbed to $344.3 million, the highest in five years and $10.7 million above 1963. B&O's increase of 3.5 percent in carloadings was one of the better showings amount all Eastern railroads."

* * * * *

"B&O's coal revenues-$102.6 million-were the highest since 1957 and $1.5 million above 1963. In making this better showing, B&O trains hauled about 48.5 million tons of coal, an increase of some 4.5 million tons over 1963. Coal loaded at

B&O-served mines increased from 32.7 million tons in 1963 to 37.3 million tons. Of all types of movements, shipments to Great Lakes ports showed the heaviest increase-2.2 million tons."

"Continuing industrial development in B&O territory promises to increase revenues in the years ahead. A total of 180 new industrial projects were begun along B&O's lines in 1964, representing a total investment of $40.5 million."

* * * * *

"Did you ever hear applause on a train? B&O trainmen do every night at the end of the free movies shown on the *Capitol* between Baltimore/Washington and Chicago, and the *National* between Baltimore/Washington and St. Louis."

* * * * *

"Also introduced were Red Circle Day Fares which reduce one-way coach and first-class tickets by 31 percent on Monday, Tuesday and Wednesday. In many cases, the lower fares make the train the least expensive way to travel."

* * * * *

"An outstanding compliment was paid to The Greenbrier through its Executive Food Director, Herman Rusch. In 1964, Mr. Rusch served as Food Director for the American and British teams at the Summer Olympic Games in Tokyo. Since 1957, he had held the same assignment for all American Olympic Teams."

* * * * *

Lyndon Johnson was elected President with a promise to create the great society. He escalated the Vietnam War, but got completely discouraged before his term was over.

My good friend, Dave Zegcer, who was in charge of the Bethlehem mines at Jenkins, Kentucky, notified me that a new mine was in the planning stage. Jack Davis who was our representative in the field at that time accompanied me with members of the operating and engineering department to the site about five miles east of Jenkins, Kentucky. After the meeting, we went back to Dave's office at Jenkins where he furnished us with maps and reserve studies to support the new mine. Our report to Cleveland was positive in every aspect. Since the B&O merger, it had been decided that the operator should be required to advance the cost prior to construction and the money would be refunded at a fixed sum for each loaded car shipped from the mine.

Dave was a stickler for appearance and after the tracks were completed, he asked me to have our company seed the slopes along the new tracks to cover them with grass. I had to tell him that I didn't even have the nerve to forward such a request.

I had been advocating the hiring of a graduate mining engineer for our office and finally received permission to proceed. After surveying the list of available graduates, we decided to offer the job to a young man by the name of Steven Rapchak at Penn State. Jim Lilly and I met with him the morning of the annual football game between WVU and the Lions that afternoon at State College. The game proved to be a disaster to the tune of 60-6 and we got rained on. My brother later joined the staff there and I never could bring myself to accept his invitation to see another match.

One of our clerks, Linda Ashcraft, decided to retire and stay home to raise her children. We therefore brought in a new replacement by the name of Jane Anderson. She proved to be a good worker with an outgoing personality who made herself very helpful in our small office.

Several management promotions were announced along with consolidation of most departments for better operating control and efficiencies. Greg DeVine was elevated to President of C&O and Executive Vice President of B&O. Hewitt Biaett from the Law Department had become a good friend of DeVine and was appointed Vice President of Coal Traffic, for the B&O with an office in Baltimore. For the lack of a Vice President of Coal Traffic and Development for C&O, George Sandmann became the coordinator in residence on the twenty-first floor of the Terminal Tower.

C. Vernon Cowen became Vice President of Operations for both roads upon the retirement of Irvin Dunn. He had worked his way up in the department from brakeman and meager education. John P. Hanley of C&O became Vice President of the Tax Department for both roads. M. C. Mulligan of the B&O became Vice President of Real Estate for both, Robert McGowan of C&O was made Vice President of Planning, Hays T. Watkins of C&O was appointed Vice President of Finance and Treasurer for both, John T. Ford of C&O was appointed Comptroller of both and Paul A. Riestrip of B&O was appointed Director of Passenger Service for both roads.

My own personal contribution took a big step forward with my successful idea for a big relocation project in eastern Kentucky but my family life was severely disrupted by the discovery of incurable breast cancer in my wife in July.

One of the parts of the Big Sandy canalization plan included the development of a large impoundment dam on Levise Fork south of Pikeville. It was intended to supply water for the operations of the canal during the summer dry season and the local populations of Pike County continued to promote it for purposes of recreation and flood control. The C&O had a branch line along one side of the river in the area and our management insisted that the rail line be preserved even if necessary to raise the tracks to a higher elevation and provide a tunnel with flood gates through the dam. Such an arrangement was very much opposed by the local populace who considered the rail line an eyesore and impediment to recreational uses. The C&O was interested in maintaining access to the coal reserves in the area though there was currently only two small truck mines operating there.

On one of my field trips visiting with Gates Engineering, I explained our problem with the project and John Straton an officer in that company suggested that we might try relocating the rail line around the area of the dam and impoundment.

With this idea, I went back to our office and assigned Jim Lilly to drop everything else and develop a reserve study of the Johns Creek area immediately north and parallel to Levisa Fork. In a week he came back with a map and study showing much more and better coal reserves in the Johns Creek watershed. I wrote a complete report and recommendation to relocate our rail line over the dividing mountain into Johns Creek and to abandon the existing line which was interfering with the construction of the proposed dam. My recommendation was adopted by DeVine and the Federal Government agreed to provide $11 million for the relocation. The line was built with heavy grades and curves which limited its efficient operations but several mines opened immediately and greatly increased current coal traffic with every indication that more development would ensue. The operations and engineering departments took a dim view of this project because of severe grades and curves required. Their preference was to take the $11 million and forget about the coal reserves.

While all of this was transpiring, I had previously been serving on the School of Mines Advisory Board at WVU. In addition I was elected President of the West Virginia Society of Professional Engineers in July and Chairman of the Appalachian Section of the American Institute of Mining Engineers in the fall.

My family and professional life was vastly complicated by my concern and effort to help my wife cope with her distress. My son was a senior in high school that fall and getting ready to attend college the next year.

C. P. Anderson had been a successful coal operator on the Virginian Railway succeeding his father, but fell on hard times. He came to see me looking for a job and was hired to be a field representative and later transferred to the B&O. C. P. had a neighbor by the name of Jack Long who operated a manufacturing company started by his father. He was a very talented mechanical inventor and successful businessman.

I learned that the company had given up on trying to sell the Chafin building and were going to sell the one we were in. The operating headquarters had been moved to Baltimore vacating the first floor of the huge OH building located in the west end of Huntington. A representative from the real estate department came to see me and advised me that our office would have to be moved to the OH building. The prospect of moving out of our quarters was very distasteful to me because it was in a very convenient location downtown and we were all attached to it. However, I knew the die had been cast and agreed to look at a suggested suite of offices in the OH building. What I saw was an area without any windows that had been briefly occupied by the Real Estate Department which seemed very depressing to me. I made my objection to the area very clear and wrote a memo to the Cleveland office about my objection to this location. A couple of weeks later, I was advised that we would move to the former operating headquarters.

office in front of the OH building. I knew from previous visits that this was the best company offices in Huntington and readily agreed to move in. I was authorized to buy some new furniture and we were soon back in business. The moving of our files, books and maps was a horrendous job and we discarded two truck loads of obsolete records in the process.

C&O DIRECTORS

Cyrus S. Eaton
Walter J. Tuohy
Gregory S. DeVine
Robert J. Bulkley
Barnum L. Colton
Cyrus S. Eaton, Jr.
Roger H. Ferger
M. S. Fotheringham
John E. Kusik
Dr. Fay A. LeFevre
Erwin H. Hill

B&O DIRECTORS

Walter J. Tuohy
Gregory S. DeVine
John D. Biggers
Cyrus S. Eaton
Milton S. Eisenhower
Charles S. Garland
Edmund L. Grimes
John E. Kusik
John E. Motz
James Parker Nolan
Howard E. Simpson
Arthur W. Steudel

OFFICERS OF C&O AND B&O

William R. Althans .... *General Counsel-C&O*

Hewitt Biaett .... *Vice President, B&O, Coal Traffic & Development*

Owen Clarke ..... *Vice President, C&O-B&O, Personnel, Labor Relations, Purchases and Materials*

C. Vernon Cowan .... *Vice President, C&O-B&O, Operations*

James E. Doyle .... *Vice President, C&O, Merchandise Traffic*

Kenneth E. Ekin .... *General Counsel, B&O*

John P. Ganley .... *Vice President, C&O-B&O, Taxes*

Walter M. Haenssel .... *Vice President, B&O, Freight Sales*

Charles Henry, Jr. .... *Vice President, B&O, Marketing*

M. C. Mulligan .... *Asst. Vice President, C&O-B&O, Real Estate*

Robert C. McGowan .... *Vice President, C&O-B&O, Planning*

Howard Skidmore .... *Vice President, C&O-B&O, Public Relations and Advertising*

Hays T. Watkins, Jr. .... *Vice President, C&O-B&O, Finance (also Treasurer, C&O)*

E. W. Wright .... *General Manager, Trailer Services, C&O-B&O*

T. Howard Keelor .... *Corporate Secretayr, C&O*

August W. Schell .... *Corporate Secretary, B&O*

John T. Ford .... *Comptroller, C&O-B&O*

Edward P. Snyder .... *Treasurer, B&O*

Paul H. Reistrup .... *Director of Passenger Services, B&O*

☆

The former Operating Headquarters Building on Madison Avenue in West Huntington.

The C&O Building located at the corner of Fourth Avenue and Eleventh Street in downtown Huntington.

1965 - MERGER WITH N&W PURSUED

The letter format started out with the merger news:

"Biggest news of the year-of many years-for C&O/B&O was the agreement for merger of C&O with the Norfolk and Western Railway. We believe this is the most important move for greater service growth and strength that could be made."

* * * * *

"C&O's consolidated earnings rose to $65 million, equal to $7.45 a common share, compared with $50 million, or $5.80, in 1964. The consolidated earnings include C&O's own net income, plus its 90 percent ownership of B&O and C&O/B&O's 65 percent ownership of Western Maryland Railway."

"The $4 dividend was paid by C&O for the ninth consecutive year. Since that rate was set, $305 million has been paid in cash to shareowners of C&O."

"As reported to you at the time, C&O's Board of Directors on August 31 approved, as did the N&W's Board, the proposed merger. Formal application to merge was made to the Interstate Commerce Commission on October 11."

"In the merger, each C&O common share would be exchanged for .725 a share of N&W common."

* * * * *

"C&O shareowners will have an opportunity to vote on the plan at the C&O Annual Meeting, which this year will be held in Cleveland on May 12."

"B&O would, of course, be part of the new system, as would Western Maryland if the ICC approves C&O/B&O's pending application for control. N&W-C&O also would accept as a condition to approval of their merger, on specified terms and through subsidiary companies, and ICC requirement that they offer to include in their system five other railroads. They are the Delaware and Hudson, Erie-Lackawanna, Boston and Maine, Reading and Central of New Jersey."

"The N&W-C&O merger plan would create a large, strong system connecting most of the major centers in the northwestern quarter of the nation. It would be about equal in gross revenues, assets and miles of road-and would provide true competitive balance-with the proposed Penn-Central system, which has already been recommended for approval by the ICC hearing examiners. Summing up, it would make possible a new era in railroading in the East."

* * * * *

"C&O/B&O outlays for equipment and improvements totaled $90 million in the year. To assure continuing strength for C&O/B&O, it is planned to spend $175 million in 1966 for new locomotives, more freight cars, data processing equipment, electronic communications systems and other improvements."

"On July 27, Examiner Paul C. Albus of the Interstate Commerce Commission recommended that C&O/B&O's application for control of the Western Maryland be granted. He stated that inclusion in the C&O/B&O family of the 844-mile freight railroad was 'logical and desirable.' Now final approval of the Commission is hopefully awaited."

"Earlier in the year, C&O applied to the ICC for control of the Chicago South Shore and South Bend Railroad, a small but important line serving a rapidly expanding industrial region on Lake Michigan. Hearings on this case will begin January 10 in Washington."

* * * * *

Results for Each Railroad (in millions of dollars)	C&O		B&O	
Operating Revenues	1965	1964	1965	1964
Coal and Coke	$180	$175	$103	$103
Merchandise	190	173	261	242
Other	27	25	36	37
Total Operating Revenues	397	373	400	382
Expenses, Taxes, etc.-Net	349	331	384	375
Net Income	48	42	16	7
Working Capital	46	40	23	(9)
Cash and Short-Term Investments	72	67	45	29

"Coal is carried in greater tonnage by C&O/B&O than any other commodity. Of the 121 million tons of coal C&O/B&O moved in 1965, more than three-quarters came from mines along the system's lines."

* * * * *

"One ton of coal can produce 1,500 pounds of coke for the blast furnace, ten gallons of coal tar for bug-killers and drugs, three-and-a-half gallons of light oil for nylon, synthetic rubber and TNT, and twenty-four pounds of various chemicals."

* * * * *

"However, they grow, grains and grain products are good business for C&O/B&O-4.1 million tons in 1965 and $14.6 million in revenues."

* * * * *

"Pennsylvania and New York Central have a merger application pending before the ICC. This Penn-Central system, because of its size, resources and revenues, would, in the opinion of the Boards of Directors of N&W and C&O, eventually dominate the existing Eastern rail market to the detriment of N&W and C&O unless the proposed N&W-C&O system is created. Only with such a N&W/C&O system can a true competitive balance be achieved in the East."

* * * * *

	N&W-C&O	Penn-Central
Gross Revenues	$1.9 billion	$1.9 billion
Assets	$5.9 billion	$6.0 billion
Miles of Roads	26,460	23,271

"The five railroads proposed to be included in the N&W-C&O system through subsidiaries (tentatively called the Dereco lines), without assumption of their indebtedness by N&W-C&O, had total gross revenues of $494.2 million and total assets of over $1.5 billion in 1965."

* * * * *

"Robert J. Bulkley, former United States Senator from Ohio, and a Director of C&O for nearly twenty years, died July 22, 1965. Mr. Bulkley was Senior Partner in the law firm of Bulkley and Butler of Cleveland."

* * * * *

On my travels in the Logan field, I became acquainted with the Jones brothers. Charles was the oldest and preferred the river operations of the family's holdings. Herbert was an engineer and stayed closer to the mine

MELVIN TRIOLO

JOHN R. FIELDS

ARNOLD R. MILLER

ALLEN S. PACK

which operated under the mantle of the Logan County Coal Corporation. Herbert always took a personal interest in every aspect of a new siding for one of their mines.

There were several important appointments during the year. Hewitt Biaett made Vice President of the Coal Traffic and Development of C&O as well as the B&O filling a gap left by Greg's elevation to President of B&O/C&O. Kenneth Elkin, of the B&O, was made Vice President-Law for both roads, B&O/C&O. E. W. Wright of B&O was made General Manager Trailer Services; Walter M. Haenssel of B&O was made Vice Prsident of Freight Sales for B&O and Charles Henry, Jr. became Vice President Commercial Development for both roads.

The hearings before the ICC to get approval for the latest merger were held later in the year and my opposite member, Douglas Krickmer, of the N&W was selected to prepare testimony regarding coal development for both roads. I knew from this turn of events that he would take over coal development in the new company which was a hard pill to swallow on top of the fact that I had not even been approached to take over the coal development function on the B&O. Just before the B&O merger, that company had hastily set up its own coal development office in the First National Bank in Fairmont, West Virginia. Harold Beatty was hired to head it up. He was a Pitt graduate in mining and was General Manager for the Elk Creek Coal Company at Widen, West Virginia, in 1952 during the strike by the UMWA to organize the mine. He tried to keep the mine working during a 15 month reign of terror.

During all of the year, my home life was one long nightmare due to the gradual worsening of the cancer my wife was fighting. Fortunately, her folks were staying with us most of the time which permitted me to keep up my routine of travel in the coal fields. She insisted that our son go away to WVU even though her life was ebbing away. She even tried to console me about my disappointment in not having my job expanded to include the B&O.

During the year, I started reporting to Hewitt who had moved his office from Baltimore to Greg's old office in the Terminal Tower in Cleveland. I found him to be an easy going, intelligent and sensitive individual who was most considerate of his fellow workers. He also seemed to be under a lot of stress in his job running the coal department and I assumed it was because he was operating in a new area away from the law department.

By this time, I was spending a lot of time traveling with Biaett and DeVine as an advisor in their meetings with leading coal operators and other important people. I had become used to traveling first class and flew a lot in the company planes. We always stayed at the best hotels and ate in the best restaurants.

Ever since R. O. Robertson retired as head of the Real Estate Office in Huntington, I had been trying to manage the timber resources on our coal lands in a manner that would make money and provide for future growth. At the invitation of J. P. Hamer who was one of our largest timber contractors, I joined the American Hardwood Association which was a trade group interested in promoting the hardwood lumber industry. I wanted to do a good job with our timber resources and realized that we needed a graduate forester to do a proper job. After several requests to DeVine and references to the need in our correspondence, he authorized me to hire someone. After making several inquiries about someone for the position, a small timber contractor in Huntington called me on the phone and highly recommended a young man who was working for the State of Kentucky. I contacted him and asked him to come in for an interview.

Robert Forney arrived in my office on the day after my wife, Nadine, came home after her breast cancer surgery. I invited him to the house after work to meet my family. As soon as he learned of my wife's experience, he said, 'I know just how you feel, I had the same operation when I was in high school.' I was so surprised at this statement, that I nearly fell off of the bannister on the back porch. He then explained that he was cleated in the chest playing football and the lime used to line the field got in the wound and caused an infection. His breast was then removed to heal the wound. After he was hired, Forney negotiated a timber contract at prices nearly double the last one I had made. He was such a good employee that I developed more respect for Penn State.

It seems to always happen when you acquire a good man you lose one and in this case, we lost two. Dick Barnett had always made a good impression on the people in Cleveland and they asked him to fill a vacancy in the sales office in Detroit. I hated to lose him but wanted him to get ahead so I encouraged the move.

J. B. Herring reached the retirement age of 62 and left a big hole in our office organization. He was in perfect health at the time and I felt that in his case, it was premature.

☆

1966 - GRIM REAPER PUTS DeVINE IN CONTROL

The annual report featured a memorial to Walter J. Tuohy, who succumbed to another heart attack.

"Walter Tuohy's death on May 12, 1966, brought to a close a distinguished career of more than twenty-three years with the Chesapeake and Ohio Railway Company and The Baltimore and Ohio Railroad Company."

* * * * *

"A soft-spoken man of many talents, his accomplishments were countless. He will long be remembered for having single-handedly engineered the affiliation with the Baltimore and Ohio and for the key role he played in the pending merger with the Norfolk and Western."

"But he will be remembered best for the man himself and for his genuine concern for his fellow man. At the very moment he was stricken, his thoughts were of people-the thousands who would overwhelmingly support the merger-not the millions of shares they would vote. It was fitting that the last word he wrote was 'individuals."

"Quick and wiry, his vitality knew no bounds. Happily for those whose good fortune it was to have worked with him. He possessed that uncommon ability to ease the tension of those around him. They will most certainly miss his buoyant presence and skillful guiding hand. He was a gracious leader with a genius for friendship. His warmth and sincerity earned him the universal affection of those who knew him."

* * * * *

My own family had been decimated on January 13 when my wife was placed in an oxygen tent at the C&O hospital in Huntington and I was called while on a trip to Cleveland headquarters and advised that death was near at hand. Mr. Biaett immediately called me out of a meeting and told me that a company plane would be waiting for me at the airport. I arrived at her bedside about five hours before she expired.

The annual report letter was signed by Cyrus and Greg who immediately succeeded Walter as CEO of B&O/C&O. The text proclaimed that C&O/B&O had enjoyed another good year financially and the B&O contribution was evident:

"Earnings of B&O in the year made an important contribution to C&O's consolidated net income. In 1961, B&O had suffered a deficit of $31 million. Thereafter, in partnership with C&O, its earnings rose as follows: $1.6 million in 1962, $5.5 million in 1963, $7 million in 1964, $16 million in 1965, and $22 million in 1966. The latest result shows a heartening continuation of the upward trend."

* * * * *

"C&O had a record year in merchandise freight revenues. B&O's increased volume of general freight handled produced revenues substantially higher than in 1965. New industries that located on C&O/B&O lines during the year, and present businesses that expanded, are expected to provide over $18 million in additional freight revenues annually."

"Piggyback continued climbing. From transporting highway trailers on flat cars, B&O took in $26.6 million, 27 percent better than 1965's results. C&O's revenues of $9 million were 14 percent better."

"Coal traffic tonnages, up slightly from the year before, would have been even better except for work interruptions at the mines. C&O revenues were affected by a decrease in overseas export coal and shifts in traffic patterns. Development programs for future coal traffic moved ahead. For example, contracts were signed under which C&O/B&O will move forty million tons of coal over the next twenty years to two new electric power units. Major branch lines were completed to tap vast coal deposits on both C&O and B&O, and plans were made to start other branches to serve the new coal operations in 1967."

"Net income of C&O, reflecting the 1966 coal revenue situation, was $45 million. This compared with $48 million the year before, but was still the second best C&O net income in the last seven years. B&O net income of $22 million, was up $6 million."

* * * * *

CONSOLIDATED EARNINGS	**1966**	**1965**
Consolidated Net Income (millions)	$66.1	$64.7
Consolidated Net Income per C&O Common Share	$7.61	$7.45
Combined C&O/B&O Results		
Operating Revenues (in millions of dollars)		
Merchandise	470	451
Coal and Coke	279	283
Other	64	62
Total Operating Revenues	813	796
Expenses, Taxes, etc.-Net	747	731
Working Capital	65	69
Cash and Short-Term Investments	122	117

"The Interstate Commerce Commission on December 30 announced that hearings on C&O's proposed merger with the Norfolk and Western Railway will start April 17 and will be conducted by Examiner Edward L. Boisseree. The vital necessity for this merger, already voted on favorably by the shareowners of both companies, was heightened by ICC approval in April of the Penn-Central system."

* * * * *

"Western Maryland Railway hopefully will become a partner of C&O/B&O in the new year. Affiliation of the 844-mile freight railroad with C&O/B&O was recommended by an ICC examiner on July 27, 1965. A final decision by the Commission is awaited."

"Chicago South Shore and South Bend Railroad today became affiliated with C&O/B&O under a November 14 order of the Commission authorizing C&O control of South Shore. This short but important line serves its service to the vast industrial complex rising around Burns Harbor, near Gary, Indiana, where Bethlehem Steel and National Steel are putting up new plants."

"Assuming a continuation of the nation's high levels of business generally, C&O/B&O anticipate increased revenues in 1967."

* * * * *

"In these efforts, as in the past, they will rely on the men and women who are C&O/B&O. In an increasingly mechanized world, the two railroads still recruit 3,000 men and women a year."

* * * * *

"At the time, the Board of C&O consisted of nine men and the B&O Board had 12."

C&0 Board	B&O Board
DeVine	DeVine
Eaton	Eaton
Kusik	Kusik
Eaton, Jr.	Motz
Fotheringham	Rentzel
Will	Steudal
Ferger	Nolan
LeFevre	Simpson
Colton	Biggers
	Garland
	Eisenhower
	Grimes

* * * * *

Several management changes were made during the year. Charles A. Campbell succeeded James Doyle as Vice President, Merchandise Freight Sales of C&O; William L. Ollerhead, Vice President Industrial Development and C. R. Zarfosse, Vice President Merchandise Sales, B&O. A picture of Hewitt Biaett also included the information that he is a native of Arkansas; earned a law degree at the University of Arizona and had previously worked as a commerce attorney.

Greg DeVine, Hewitt Biaett and several other associates from Cleveland attended my wife's funeral on Saturday, January 15, in Huntington. I was totally unprepared for the personal devastation I felt at my loss. Our only son was in shock after the funeral but returned to school. I was called back to Cleveland the next weekend and advised by Hewitt that I was to take over the B&O Coal Development office immediately and that this move had been delayed because of my personal situation. I told him that I appreciated his concern, but would have preferred that there had been no delay during which time I had concluded that I had been passed over.

The next week I spent at the B&O office in Fairmont where an office had already been prepared for my arrival. Another week was spent on a trip to Baltimore where I met the Coal Traffic people in the new office building that the company had purchased to supplement the lack of space in the old B&O building which resembled a national monument. Here I met Ross Conlon who became a dear friend over the next few years and a crew of good people who accepted me with open arms.

I spent a lot of time over the next few months traveling over the B&O coal territory with my new associates. In may ways, the experience was helpful to me in my grief-stricken mental state. I automatically increased my intake of grain spirits to ease the pain. I could only sleep about four hours a night which reduced my alertness during the day, but I never missed a day or an assignment and gradually began to get back to my old normal routine, but it was never the same again.

The annual New River operators golf outing was held as usual near the end of June at the Greenbrier. I was greatly flattered when Hewitt asked me to play with him and two coal operators. The competition must have inspired me because I shot a 90 for the best score ever at the Greenbrier. When we checked in at the men's dressing room to change clothes, we learned that Greg had developed indigestion while playing and was sitting on a bench with a towel draped over his head. I told Hewitt that we should seek out a physician to check on him, but Hewitt demurred. After several minutes went by, I suggested to Henshaw that he get a doctor for him since he was such a good friend but he also demurred. I then looked around the area and found my doctor from the clinic, Gene Morhous. I asked him to go over and check him out and Gene said he wouldn't dare do anything unless requested by Greg. I became slightly frantic by this time thinking that he may be having a heart attack and noticed that he had put on a shirt and pants and told the attendant he was going back up to the hotel. I decided then to call the clinic on a house phone. Dr. James Baker, who ran the clinic was on duty and answered the phone. I told him that Greg was ill and was coming up to the hotel in the limousine shuttle. He told me that he would meet him when he got to the hotel and knew just how to handle the situation.

Dr. Baker arranged for two heart specialists from Cleveland to fly down to the Greenbrier on Saturday evening. They accompanied him back to Cleveland with Greg on a stretcher in the back of the plane. On landing, he was immediately transported by ambulance to Charity Hospital.

Later I learned that Greg had been flown back to Cleveland on the company plane on Sunday afternoon. Several days later, it was announced in Cleveland that Greg had suffered a mild cardiac distress and would be hospitalized for several weeks. I don't think Greg ever learned who probably saved his life at the Greenbrier.

A couple of months later, I met Greg and Hewitt in Washington and accompanied them to a meeting with an official in the Interior Department. The meeting lasted a couple of hours and when it was over Greg said he was taking a cab to his hotel and would see us the next day. Hewitt rushed me down the street and into the first bar we came to. He ordered a double martini and I had my usual Manhattan. He told me that Greg was a changed man since his illness. He was taken off spirits completely and was now trying to keep everyone under him from drinking. He no longer looked his old buoyant self. Hewitt then told me he was going to make four new assistant vice presidents in the department and I would be one of them. This news was so exhilarating and surprising to me that I was nearly speechless.

When I first came with the C&O, a new branch line was under construction to cross over the mountain separating the State of Kentucky and Virginia to serve a new projected Clinchfield Coal Company mine designed to produce a million tons a year near Pound, Virginia. This was a very costly project and required a 200 foot tunnel near the top of the mountain. At the time, the railway policy was to build the tracks required for a new mine if the company and mine was judged to be the first class business venture by the Coal Development Office. In this case everything seemed to be in order and the line was constructed at a cost to the company of several million dollars.

Production at the mine never did reach the projected quantity and the underground workings were connected eventually with a mine served by the Clinchfield railroad. The company notified us that most of the remaining reserves would be diverted to other mines due to market conditions. Our management then decided to abandon the branch line which required ICC approval because two states were involved and the people in Wise County were adamant that the line be kept intact for a couple of small truck mines and hopefully for projected future business.

A hearing was held in June at the Courthouse in Wise, Virginia. I personally attended with one of my assistants, Frank Gaddy, and two company lawyers from Cleveland handled our cases, Doyle Morris and Garth Griffith. We testified with maps and other information that the remaining reserved in the area would never support the line and its maintenance costs. The local people streamed into the hearing to convince the examiner that the Line should be retained for employment possibilities. I don't think I ever spent a more miserable two weeks in my career.

The weather was sultry hot, the courtroom benches were hard as a rock and there was no air conditioning. My depression and discomfort was intensified by the fact that the examiner ruled against us but he was later overturned on appeal. The line was finally abandoned six months later and the track materials salvaged for other use.

Later in the year while on one of my flying trips to Cleveland to report to Biaett, the Vice President in Cleveland, he invited me to go home with him for dinner. I decided early in my career that I would much rather eat by myself in most cases rather than go to someone's home. I would have demurred if he hadn't been my boss. When we got to his house, his wife informed us that we were going out to eat. During the evening it became obvious that they were not getting along very well.

My steady girl friend, Claire Sheets, was teaching that year in Lakewood, a suburb of Cleveland. On one of my trips to Cleveland, she met me at the airport. It was a coincidence that Greg was deboarding the company plane at the same ramp as my flight on Northwest. I introduced her to Greg and we walked together to the main terminal. I suspected by his actions during this interlude that he was very cool towards our relationship.

When I took over the development office on the B&O, I was really returning to my roots having been born and reared in the small mining town of Erie only eighteen miles south of Fairmont. The office was on the seventh floor (the same as our office in Huntington) of the First National Bank. It had previously served as the division headquarters of Consolidation Coal Company, the largest coal company on the B&O. Harold Beatty occupied the main corner office but he had fixed up an even larger office next door to me.

It was obvious from the beginning that Beatty resented my intrusion in his domain but the other members of the office were most cordial and helpful. Rex Jones was the most effective member of the staff. He worked in the coal fields around Summersville, West Virginia. He was a little younger than I and had been working on embassy staffs in South America before joining the B&O. Harry Fischer was an elderly man transferred over from the operating department who was very loquacious and handled sidetrack matters in northern West Virginia. He volunteered information about a coal property near Phillipi, West Virginia, that had once belonged to U. S. Steel. He was rapidly reaching retirement age, so I only worked with him a few months.

Several months later, a vice president of Republic Steel called me on the phone and requested suggestions about a viable coal property on the B&O. I told him about the property Harry had described. Nothing more was ever said until a year later when Republic requested a meeting on the ground to discuss sidetrack for their newly developed mine on the suggested property.

Joe White was a former clerk who was the field representative in Pennsylvania. He was more reserved than the others but got the job done. Tom Tonkery, Chief Clerk in the office was a rabid WVU fan with whom I could trade observations and "Bernie" Juric was a female clerk who was dedicated and efficient.

One of the hardest features of my job in administering the coal development office was getting raises in salary for my staff and myself. I felt a strong obligation to see that my staff was fairly compensated for their efforts in helping me keep up with all of our obligations and responsibilities. I had no access to the other salaries in the company, but I had a pretty good feel for what my engineering colleagues were getting which I needed to back up my presentations to management. I also knew that the only way a large company like C&O could make profits was to hold down expenses which was the 'bear' I had to wrestle with. One factor that spurred me on was my convictions that there were quite a few Vice Presidents making a lot less contribution to income than my own office.

After Greg moved up to the President's office on the thirty-sixth floor, I had to go up there to confer with him about the major decisions pending in our office. On one of these trips, I prepared a presentation for raises for several members of my staff that I felt were long overdue and required redress. I waited until after we had discussed all the problems pending from the pile of files on his desk. I waded into my speech about raises. It was late in the afternoon and it was obvious that he had been over some rough spots during the day, but I was totally unprepared for his violent reaction. As soon as I had gone far enough that he understood what my pitch was, he turned red in the face, pounded his desk and got up in an obvious belligerent attitude. I don't remember now what he said, but I quickly got the idea that he was furious. Needless to say, I backed off from further discussion and made a hasty retreat towards the elevator. Evidently I had impinged upon an extremely sensitive nerve at the wrong time. I didn't let the incident restrain me from future efforts, but I did become more selective about the time and place.

☆

1967 - A VIGOROUS AUSTERITY PROGRAM

The flash annual report published on January 2, 1968 was reproduced in the annual report as follows on what was referred to as a "mixed year."

"Revenues held at reasonably satisfactory levels, in spite of the slower pace of the economy and nation-wide work stoppage on the railroads and in major industries."

"On the other hand, the costs of doing business, particularly wage rates, rose sharply."

"As a result, consolidated net income was down, in spite of what was accomplished in a vigorous cost-control program."

"C&O's annual dividend of $4 was paid for the eleventh consecutive year. At the same time, future earning power was strengthened by substantial expenditures to build up the C&O/B&O."

"This expansion and improvement was largely in new equipment and facilities to move even faster and more economically the thousand and one commodities that C&O/B&O customers ship over our rails. Outstanding among the improvements, toward which $179 million was paid in the year, were the following:"

"An $8 million pier, larger than five football fields, for transferring merchandise cargoes between trains and ships was dedicated on October 12 at C&O's great port at Newport News, Virginia."

* * * * *

"At C&O's ocean terminal and at the Locust Point Terminal in Baltimore served by B&O 23,000 vessels from ports all over the world loaded and unloaded nearly one million tons of general cargo, exclusive of coal, ores and grain."

"A $3.5 million piggyback terminal at Forest Hill Yard in Chicago started operations in November."

* * * * *

"C&O/B&O piggyback revenues have grown to ten times those of 1959. Over eight percent of total merchandise revenues in 1967 was derived from trailer services. It would take 200 trucks and 200 drivers to move on a highway as much freight as a 100-car piggyback train moves on the rails."

"At a $1.2 million automobile preparation terminal at Jessup, Maryland, automobiles fresh from far-away factories roll off tri-level railroad rack cars."

* * * * *

"Three of the leading car makers-Ford, Buick and Oldsmobile-use the Jessup terminal to deliver new models to the Washington-Baltimore area. The 65-acre terminal can store up to 2,000 autos and trucks. The railroad industry's share of the number of automobiles moving to market has zoomed from eight percent to 50 percent in less than a decade. Revenues have grown to $300 million annually."

"The weapon that won the auto traffic away from the haulaway trucks is the rack car, simply a longer flat car with one or two extra decks, ramps to load twelve to fifteen autos and devices to tie them in place. It produces a shipping cost per auto lower than that for all-highway movement. C&O/B&O have 1,031 of these rack cars."

* * * * *

CONSOLIDATED EARNINGS	1967	1966
Consolidated Net Income (millions)	$52.3	$66.1
Consolidated Net Income per C&O Common Share	$6.02	$7.61
Combined C&O/B&O Results		
Operating Revenues (in millions of dollars)		
Merchandise	469	470
Coal and Coke	283	279
Other	57	64
Total Operating Revenues	809	813
Expenses, Taxes, etc.-Net	757	747
Working Capital	41	65
Cash and Short-Term Investments	95	122

"A $10 million B&O pier for loading coal from trains into ships and barges was in the final stages of construction at year's end at Curtis Bay in Baltimore."

* * * * *

"While overseas export of coal has come down in volume from the high levels of the late 1950's. Heavy tonnages still move. In the last five years, from C&O's huge terminal at Newport News, Virginia, and from B&O's Curtis Bay Facility, nearly 60 million tons of coal have been shipped to foreign markets."

"For carrying everything from canned milk to coils of steel, 5,200 merchandise freight cars, costing $77 million, joined the C&O/B&O fleet."

* * * * *

"At the end of 1967, C&O/B&O had a merchandise freight fleet of 62,000 cars with a replacement value of $1 billion. These cars are in addition to coal-carrying cars."

* * * * *

"Advanced design was also apparent in the 1,900 coal hopper cars bought. All hold 100 tons, double the standard, and their wheels turn easier on roller bearings."

* * * * *

"C&O/B&O had 79,000 coal hopper cars, one-fifth of all the coal cars operated by the nation's Class I railroads. Most of the new 100-ton hopper cars were built in C&O/B&O's shops, turned out at the rate of twenty-four a day."

"Increased power to pull and push the long trains over the road or a few cars at a times in terminals came from 128 diesel locomotives acquired."

* * * * *

"Diesel locomotive economy is one reason why railroads charge only an average of 1-1/4 cents to move one ton one mile, compared with 6.5 cents for trucks and 21.7 cents for air cargo."

* * * * *

"Heavier cars and stronger locomotives need heftier track, bridges and signals."

* * * * *

"Nearly 2,300 miles of welded rail are now in used on C&O/B&O's main lines. Welded rail does more than banish the 'clickety-clack'-it lasts longer, has better riding quality and costs less to keep up. C&O/B&O spent $12 million in 1967 to improve 3,354 miles of track and to maintain the entire 11,000 miles of track they operate in twelve states and in the province of Ontario. A $1.1 million program was begun in 1967 to upgrade bridges and tracks on 377 miles of B&O line through Indiana and Illinois."

* * * * *

"These 1967 improvements brought total capital expenditures since 1962, when the joint C&O/B&O development program began, to $838 million. This money provided, along with other things, 70,000 freight cars, 604 locomotives, 2,300 miles of welded rail and a system of electronic communication, data-handling and computerization that is one of the most advanced in the industry."

* * * * *

"Coal revenues rose by $4 million, as national coal production increased by three percent to 560 million tons."

* * * * *

"Eleven new mines went into production on C&O and B&O lines, and eighteen mines that had been idle for a number of years resumed operations."

* * * * *

"Movements of coal in trainload volume to electric utilities, steel mills and other users continued to grow, with 44 percent of all coal mined in C&O/B&O territory going to market in this manner."

* * * * *

"Merchandise freight revenues almost equaled 1966, even though the economy dipped from its previous high levels and production was interrupted by work stoppages in several major industries. Over half of all the commodities moved by C&O/B&O come from factories, farms and forests in the two railroads' own territory, the rest from connections with other railroads. The biggest portion of this business is motor vehicles and parts, chemicals and explosives, metals, gravel, sand and stone, and ores."

* * * * *

"Contributing somewhat to both merchandise and coal freight revenues was a rate increase, averaging three percent, provisionally granted to all railroads in August."

* * * * *

"While the new rates will produce an estimated $300 million in additional revenues, they merely allow the railroad industry to catch up partially with these soaring costs."

"New industries that located on C&O/B&O and existing industries that expanded will provide in the future additional freight revenues of $18.5 million annually. Among the largest of the plants were three major automobile factories, two in Michigan and one in Ohio."

* * * * *

"Piggyback traffic, a steadily growing part of the railroad business, produced revenues of $38 million, up from $35 million in 1966."

* * * * *

"Passenger train and miscellaneous revenues declined by $7 million. Of this reduction, $6 million resulted from fewer riders (although C&O/B&O conducted a vigorous promotional and advertising campaign) and the loss of mail which the Post Office diverted to trucks and airplanes."

* * * * *

"Net income in this mixed year, as a result, was $52.3 million. This was far less than in 1966 and 1965, but higher than in any of the years 1960 through 1964."

* * * * *

"On January 15, the Supreme Court of the United States upheld the ICC's approval of the merger of the Pennsylvania Railroad and the New York Central System, opening the way to consummation on February 1 of this largest railroad consolidation in history."

* * * * *

"You already have been told of the far-reaching benefits, both to you and the public, expected from the proposed merger with the Norfolk and Western Railway. Steady progress was made in the necessarily length regulatory proceedings that accompany such a large railroad consolidation."

* * * * *

During the spring of the year after struggling with grief and trying to get a handle on B&O development, I received a shock which was to carry over to the end of my career. While in Cleveland I report to Hewitt about routine matters. He had taken me to the officer's luncheon in the Greenbrier suite. Harold Lane asked me later if I had heard about the new promotions and when I responded in the negative, he told me that an announcement came out that morning that Sandmann, Henshaw and Ross Conlin were promoted to Assistant Vice President. When the significance of the announcement sank in that I had been passed over I felt like I had received a blow to the solar plexus. I recognized immediately that my career had suffered a devastating wound that would be recognized throughout the Company and the coal industry and that my future executive potential was shattered.

I groped around preoccupied with my disappointment the rest of the afternoon and when the time came to head out to the airport, I made a personal decision to go back to the hotel for the night and confront Hewitt for an explanation the next morning. I didn't sleep very well and ate very little breakfast. When I arrived in the office, I stopped to tell Hewitt's secretary that I wanted to see the boss. She said he was alone and for me to go on in. Hewitt was reading something and looked up and I could tell he was surprised to see me. I sat down beside his desk and asked him why he hadn't told me about the promotion the day before. He looked at me a minute or two and said, "Greg doesn't like you, does he?" This blunt news disarmed me completely and I stammered something about the fact that I had some inkling, but didn't realize it was so obvious. I waited for him to explain further, but he turned back to his papers and said not a word. I waited about ten minutes or so and finally realized he had no intention of discussing the matter further.

FRANK L. GADDY

J. A. HUNT

JOSEPH W. LEONARD

LESLIE C. GATES

I left Cleveland that afternoon on Northwest Airline on the trip back home with a heavy heart. I knew my influence and effectiveness had been severely curtailed and I would soon find out who my friends were.

In June of that year, I marries Claire Sheets and moved her to my house in Huntington with her two small children.

I tried to visit the Fairmont Office and accompany one of the men on their rounds about once a month. I made my living headquarters in the Fairmont Motor Lodge where the lady in charge always assigned me one of the best rooms.

The movement of hauling coal in trainload volume from one mine was receiving a lot of attention at the time. Some of the railroads like the Penn-Central and Great Northern were promoting the idea of having dedicated equipment kept organized in trainload lots (6,000 tons) and used their locomotives to pull the cars through automatic loading stations at the mines. Upon completion, the same engines would haul the train to a destination such as a power plant stopping only to change crews at exchange points. The incentive for this type of haulage was that it was cheaper to operate than conventional methods and should result in a lower freight rate.

Our management contended that the system in use for many years on C&O was more efficient than unit trains. This was because the empty cars were left at each mine when the loads were pulled. The train was consolidated with others depending on whether the cars were moving east or west at the division yard such as Peach Creek yard in Logan. Long 50-100 car loaded trains were then hauled to the main eastbound yard at Hinton or the west bound yard at Russell, Kentucky. There the cars are humped and arranged in train loads for destination. Such a system was very efficient and eliminated any crew time required doing loading and unloading of the coal. There are cases that unit trains are more efficient and are being used today. When the C&O did offer unit trains, we required the mine operator to have a sidetrack arrangement and his own locomotive to load the coal cars.

On a business trip flight to Washington, D.C. during the summer, I found myself seated beside Arnold Miller, the President of the U.M.W.A. He was a very friendly seat mate and told me about living next door to Jerry West, the all-American basketball player at WVU and the L. A. Lakers, and he was forever thumping a basketball in the back yard.

It was not long before I realized that I needed assistance in taking over and managing the Fairmont Office. I asked Frank Gaddy to temporarily work out of the Fairmont office while still keeping his family in Huntington and commuting back and forth. I could tell from his reaction that he did not relish the idea. When he got out of V.P.I. he went to work for Warner Collieries on the Kanawha River and Harold Beatty was his boss. He had told me before that they never got along very well and he finally left to seek other

B&O locomotive power for mountain hauls

employment. I explained to him that I realized it would be difficult for him in Fairmont, but I would back him up. Automobile travel to Fairmont was being improved by the construction of Interstate 79, but there were a couple of gaps in the system. It was a four hour trip at the time and has been reduced to three hours now that the highway is completed.

One part of my new responsibilities that I really enjoyed was my trips to the B&O Coal Department office in Baltimore. The office was located on the 21st floor of a brand new glass and steel office building at One Charles Center which expands the B&O offices in the city. It was right next to the new Hilton Hotel in the center of the downtown area.

Ross Conlin was in charge of the office and a more positive and congenial individual could not be imagined. He was of ordinary build and proportions, but had red hair and twinkling blue eyes. One of his favorite expressions was to refer to his friends as "big hot dogs." He had a very relaxed management style and was obviously highly regarded by his staff. Two of his main assistants were Ken Bollinger and George Riley who were most helpful to me and very competent individuals. Ken Bollinger had a great sense of humor and kept the correspondence flowing while George was more introspective and a walking encyclopedia of coal rates.

We always ate lunch in a very cozy restaurant located in the basement of a nearby building where I learned how really delicious fresh seafood could be. My seafood preference was soon enlarged by the addition of Blue Point crabs and Chesapeake Bay oysters.

Ross had started out and served long in the passenger department of the B&O, but was transferred to coal when it became obvious that the railroads could no longer compete with the airlines. He had a very quick mind and had obviously learned a lot about the coal industry before I arrived. I immediately started to copy him with all of our correspondence concerning B&O matters and he reciprocated by keeping me informed of corporate news which I was not directly connected with.

I began attending most of Ross's staff meetings with his sales office people and occasionally some evidence of resentment about the C&O takeover would surface and I always countered with the statement that I would cooperate with them regardless.

Besides my very good friend, Ross Conlin, I learned to like many of his cohorts in coal traffic. Ellen Etcheson, his secretary, was a very impressive individual. Bob Grayson, his assistant, handled personnel matters. I liked all of these people and enjoyed working with them. I also got to know Bernie Dillon, Charles Schroeder and Thomas Turnbill in their Pittsburgh office.

The excellent seafood available in Baltimore certainly added to my good impression of the city. The oysters and blue crabs are in a class all by themselves.

In the summer, Greg called me and asked me to meet him at the airport in Beckley for a visit the following Friday to Slab Fork, West Virginia. Austin Caperton had invited him to have lunch at his home near the coal mine his father had opened in 1907. It was a very nice lunch which I also got to enjoy, but Greg complained to me on the return trip that he didn't like to eat meat on Friday.

☆

1968 - MERCHANDISE ALMOST DOUBLES COAL REVENUE

"Below are the comparative figures for 1968 and 1967. Comment follows each of the figures-

	1968	1967
Consolidated Net Income (millions)	$55.2	$52.3
Per C&O Share	$6.35	$6.02

"Earnings improved over the year before, reflected the divergent effect of the two main business forces at work in 1968. These were higher revenues and continuing inflationary cost increases."

	1968	1967
Merchandise Freight Revenues (millions)	$516	$469

"A generally good business year for the nation helped bring increased merchandise revenues for C&O/B&O, the highest since the two roads affiliated six years ago. Future growth in this important traffic was assured by the announcement during 1968 of major new industrial plants in C&O/B&O territory by such companies as GM, GE, Armco and Dow."

	1968	1967
Coal Freight Revenues (millions)	$276	$283

"Coal traffic, down slightly, reflected the drop in national coal production. Together, C&O/B&O transported from mines on their own lines nearly one-fifth of the nation's entire coal output. Twelve new mines were opened."

	1968	1967
Passenger Revenues (millions)	$20	$28

"As the number of riders continued to decline despite C&O/B&O's efforts to improve services and as more of the U.S. mail was taken off trains, passenger revenues dropped. The direct, out-of-pocket deficit from passenger service was $13.8 million, up from 1967 despite fewer train runs. Except for this loss, C&O consolidated net income would have been $1.12 a share greater."

	1968	1967
Total Operating Revenues (millions)	$842	$809

"The $33 million increase resulted from the higher levels of the nation's business and also from freight rate increases granted to the railroad industry."

*****

"The Commission granted an interim increase of three percent in June, and in November allowed almost the full request, on an interim basis. If finally approved, the increase will amount to $40 million yearly for C&O/B&O."

"Most trucking companies, the railroad industry's prime competitors, also raised their rates substantially in 1967 and again in 1968. The cost of shipping by rail, despite the recent increases, has actually gone down over the last ten years, while the cost of living was going up 20 percent."

	1968	1967
Expenses, Taxes-Net (millions)	$787	$757

"C&O/B&O's bill for wages, benefits and retirement taxes was $25 million greater than 1967's. To this was added higher prices for material and supplies, as well as increased depreciation and fixed charges."

	1968	1967
Working Capital (millions)	$39	$41
Cash and Short-Term Investments (millions)	$91	$95

"Financial strength remained at satisfactory levels. Working capital and cash and short-term investments at year end, although down, were entirely adequate."

* * * * *

"As the year drew to a close, C&O/B&O awaited the next steps in the steady progress toward merger of the C&O and the Norfolk and Western Railway."

"The hearings on the merger plan, after seventy-six days of sessions, were concluded in April by Examiner Edward L. Boisseree of the Interstate Commerce Commission. Shortly after, in June, important support was voiced by the Department of Justice, which told the I.C.C. that the merger 'was not objectionable from the standpoint of the public interest and, indeed, may promote it.'"

* * * * *

"C&O/B&O were authorized by the I.C.C. in March to assume control, of the Western Maryland Railway. B&O had held since 1927 a major but non-voting interest in this important 844-mile freight line operating in Maryland, Pennsylvania and West Virginia. The Penn Central and Pittsburgh and Lake Erie railroads subsequently sued to reverse the I.C.C. action, but a three-judge Federal Court unanimously upheld the Commission."

* * * * *

"Merchandise freight movements increased in several important commodity groups: iron and steel products, automobiles and automotive parts and paper and forest products. Piggyback traffic continued its excellent growth, contributing revenues of $44 million, up from $38 million the year before."

"New plant locations and expansion of present industries during 1968 will add an estimated $18 million to future annual revenues."

* * * * *

"Important traffic growth also is coming from new food distribution centers which C&O/B&O are helping develop at Fostoria, Ohio; Huntington, West Virginia, and Baltimore. These centers receive full carloads of grocery products from manufacturers and reassemble them into mixed carloads of products for supermarket outlets in several states."

* * * * *

"Coal revenues were affected by three principal factors. These were large stockpiles of coal at electric plants, lower movements of coal for overseas export and a strike at the mines in October."

"Of the twelve mines that started up in the year, the three largest will produce five million tons of coal annually when in full operation. In addition, eleven mines that had been closed were reopened, and one large mine undertook a major expansion."

"American Electric Power Company started construction near Charleston, West Virginia, of what will be the largest electric power plant in the United States."

* * * * *

"New C&O/B&O physical strength will come from the $125 million invested in capital improvements in 1968. Of this, $106 million was spent for 10,300 coal and merchandise freight cars, 850 piggyback trailers, 100 cabooses, nine diesel locomotives and 201 units of miscellaneous equipment. Another $19 million was invested in new all-welded track, new buildings, bridges and other structures and electronic systems for communication and remote control of train movements."

* * * * *

Richard Nixon was elected President confounding all the experts who considered his career ended by the defeat for Governor of California. His administration was pro business and gave the coal industry great hope.

During the late sixties I realized that we would have to beef up our engineering staff especially after Frank Gaddy resigned. I decided that we needed two young engineers to make lessee mine inspections. Since a lot of our work was in Eastern Kentucky, it would be helpful to have a graduate of UK. A young man from there by the name of Bruce Bevins was hired. He was a native of Huntington and was glad to continue living here. I went back to WVU for the other man and hired Ted Grissafi. Both of these young men were good students and fit well into our staff. They had opposite personalities, but similar goals.

It came to my attention that the State legislature had created a new set of boards to supervise higher education and it was soon to be implemented. I wrote a letter to the President of the WV Society of Professional Engineers suggesting that he recommend the appointment of one or more of our members to serve on these boards. He followed through on this matter and sent a list of five members, one from each congressional district to the Governor. My name was included on the list. Several weeks later one of our members was placed in the Regents Board which was in charge of all the state institutions of higher learning and another member, my friend Leslie Gates, placed on the WVU Advisory Board.

A couple of months later, I received a letter from Dr. James Harlow, President of WVU, that I had been appointed to serve on his Advisory Board due to the death of another member who had died before the first meeting. In this manner, I became directly involved in the affairs at WVU and tried to attend most of their monthly meetings in Morgantown. The only reward for this service was the privilege of attending a luncheon before each football game at the President's house and seats on the fifty yard line.

It was the equivalent of earning a degree in college education from my experience under the tutelage of this wise and talented individual. He was a graduate of Oklahoma University and his brother Bryce was an advisor of President Eisenhower. My tenure lasted until 1980 when I retired from the railway.

A few months after I went on the board, Dr. Harlow told us he was going to do away with the School of Mines and make it just another department in the College of Engineering. I wrote a form letter to about thirty of the most prominent coal operators and mining engineers in the State requesting that they write their objections to Dr. Harlow. In about two weeks, I got a real hot letter from Dr. Harlow complaining about the "pressure" he was receiving. At the next meeting of our boards, Charles Wise, the Chairman of our board, had convinced Dr. Harlow to retain the School which ended the problem.

Another fine resort in West Virginia is located on Cheat Lake just north of Morgantown. Lakeview Inn and Country Club was built by Ward Christopher, who inherited a coal company from his father. Many improvements have been made in recent years. My wife and I stayed there many times while attending the WVU Board Meetings.

During the fall, I came in the office on one Friday morning from a hard trip covering the coal fields in Pennsylvania when my secretary came in to give me some bad news. During the afternoon of the previous day, Bernie, the secretary in the Fairmont office called to say that Mr. Beattie had come into the office late that morning. A scene developed which unnerved all the people in the office.

On the following Monday, I learned that Harold had turned himself into the Union Hospital in Clarksburg. I discussed these matters with Rex Jones on the phone. Rex related to me other incidents leading up to this time and I concluded that I had no recourse except to have Beattie's employment terminated. I hated doing this because he was under stress due to the illness of his wife.

Early in the year, Greg advised me that the company was negotiating with Coastal States Gas Company in Texas to lease all of our available oil and gas ownership in all of our coal lands. He told me that Jack Rudnicki and Garth Griffith, one of our lawyers would work with me in the matter. His only advice to me was that I should protect all of our rights to mine coal on these properties. Griffith had been working with Rudnicki for some time on other oil and gas matters and they had become very goof friends in the process. I soon learned that they were not sold on making this lease. I explained to Greg the problem and he got Amos Bolen, one of our lawyers in Huntington, to expedite the lease. Amos quickly recognized the problem and over-rode all objections to consummate a lease in the fall of the year. All of us took an Eastern Airlines jet plane to Houston, Texas where the final negotiations were settled and we were entertained royally by our hotels. Development never did reach expectations on our lands.

C&O/B&O OFFICERS

Hewitt Biaett .... *Vice President-Coal Freight Group*
Owen Clarke .... *Vice President-Labor Relations and Personnel Group*
C. Vernon Cowan .... *Vice President-Operating Group*
Kenneth H. Ekin .... *Vice President-Law Group*
Robert C. McGowan .... *Vice President-Planning Group*
Hays T. Watkins, Jr. .... *Vice President-Administrative Group*
C. R. Zarfoss .... *Vice President-Merchandise Freight Group*
Charles W. Campbell .... *Vice President-Merchandise Freight Sales, C&O*
William J. Eck .... *Vice President-Purchases and Materials*
John P. Ganley .... *Vice President-Taxes*
Charles J. Henry, Jr. .... *Vice President-Commercial Development*
Thomas A. Keefe ...... *Vice President-Merchandise Freight Sales, B&O*
William L. Ollerhead .... *Vice President-Industrial Development*
Howard Skidmore ...... *Vice President-Public Relations and Advertising*
E. W. Wright .... *Vice President-Trailer Service*
William R. Althans .... *General Counsel*
Ross Conlin .... *Asst. Vice President-Coal Freight Sales, B&O*
H. Preston Henshaw, Jr. .... *Asst. Vice President-Coal Freight Sales, C&O*
M. C. Mulligan .... *Asst. Vice President-Real Estate Development*
Kenneth T. Reed .... *Asst. Vice President-Maintenance*
Erle T. Rucker .... *Asst. Vice President-Transportation*
George A. Sandmann .... *Asst. Vice President-Coal Commercial Planning*
David A. Watts, Jr. .... *Director of Passenger Service*
E. Truman Wright ..... *Vice President & General Manager-The Greenbrier*
T. Howard Keelor .... *Corporate Secretary*
John T. Ford .... *Asst. Vice President and Comptroller*
L. C. Roig, Jr. .... *Treasurer*

☆

1969 - WESTERN MARYLAND BOOSTS INCOME

"Revenues were a record, reflecting needed increases in freight rates. Costs unfortunately were a record also, in line with national inflationary trends and despite vigorous control efforts."

	1969	1968
Consolidated Net Income (millions)	$52.5	$53.9
Per C&O Share	$6.05	$6.21

"Consolidated net income includes C&O/B&O's equity in Western Maryland Railroad, control of which by C&O/B&O was assured in January. The combined results below include Western Maryland results."

	1969	1968
Merchandise Revenues (millions)	$563	$527

"Revenues from transporting practically everything from apples and armatures to xylophones and zinc (coal revenues are listed separately below) were up for the seventh consecutive year since C&O/B&O affiliation in 1963. The greatest gains were in shipment of automobiles and parts, chemicals, limestone and paper products."

"Piggyback shipments increased to 284,000 trailer-loads, a seven percent increase over 1968."

"Merchandise revenues for 1968, above, have been reduced to reflect a transfer of $15 million in revenues from transporting iron ore, which are now included in the coal revenues listing."

	1969	1968
Coal, Coke and Iron Ore Revenues (millions)	$310	$308

"Expectations of a substantial rise in bituminous coal revenues were dispelled in February when miners in West Virginia struck over state compensation for 'black lung' disease. This and other work stoppages later on halted or curtailed production at various mines served by C&O/B&O for over three weeks. Even though the demand for coal was very strong, and C&O/B&O operated their vast coal car fleet at capacity, not all the tonnage lost during the strike could be regained."

"The movement of coal to Tidewater for shipment overseas, mainly to japan and countries in Western Europe and South America, was up 2.6 million tons to 12.5 million."

	1969	1968
Passenger Revenues (millions)	$16	$20

"C&O/B&O have tried over the years to stop or reverse the downward trend in passenger train patronage by promoting such inducements as bargain fares and innovations in service. Despite this effort, passenger train revenues were down again in 1969. The direct loss from passenger services came to $11 million, and while this was down $2.8 million as a result of tight cost controls, it still represented a serious drain on earnings, equal to ninety cents a C&O common share."

	1969	1968
Total Operating Revenues (millions)	$924	$888

"With record revenues offset by record expenses, the C&O/B&O in October joined the railroad industry in asking the Interstate Commerce Commission for authority to raise freight rates by six percent. On November 18 the new rates were granted temporarily, pending further investigation. If finally approved, the rates would add an estimated $47 million to C&O/B&O revenues on an annual basis."

	1969	1968
Expenses and Taxes (millions)	$898	$864

"The largest single factor in higher expenses was increased payroll costs. Wages, fringe benefits and payroll taxes totaled $472 million, up $15 million from the year before."

	1969	1968
Dividends, Interest & Net Income (Millions)	$26	$30

"Non-operating properties continued to provide an important supplement to earnings generated by railroad operations."

	1969	1968
Working Capital (million)	$40	$45
Cash and Short-Term Investments (millions)	$85	$96

"Entirely adequate for the companies' needs, the levels of working capital and cash were evidence of C&O/B&O's continuing financial strength."

* * * * *

"Transportation of metal products, including steel pipe, is a $90 million business on C&O/B&O-and the pipe producers had a problem. To reduce milling costs, they were making their pipe in longer lengths-too long for most rail cars. Also, price competition made it necessary to cut delivery costs by getting more pipe into a shipment. Further, losses from pipe damage had to be reduced."

"The C&O/B&O's market and product development teams came up wholly new open-top pipe gondola cars."

* * * * *

"Hands off the Liquor. The distillers also had a problem. They shipped liquor stacked five cases high on wooden shipping platforms, or pallets. In loading boxcars by fork-lift truck, one pallet load was topped with a second pallet load-but the roof was just too low to take the top-most case."

* * * * *

"C&O/B&O specialists designed the 'Five on Five' car. They added an extra foot of height inside, but only three inches outside to stay within clearances. The inside inches came from thinner flooring and redesigned support rails."

* * * * *

"Auto-Trans is the recently developed answer by C&O/B&O's marketing and operating teams. The automobile is put aboard a multi-level rack car (the same kind used for Detroit's new models) and makes its trip undepreciated and at a cost lower than any other way. It's handed over freshly washed."

* * * * *

"Among the 285 industries which decided during the year to place new or expanded facilities on C&O/B&O were General Electric, General Motors, Chrysler, Seagram, Montgomery Ward and Anchor-Hocking. Construction started on a three-mile spur to serve GE's planned $250-million Appliance Park East at the completely new City of Columbia, Maryland, between Washington and Baltimore."

"To reach a mine opened by Consolidation Coal near Beckley, West Virginia, a nine-mile branch was built through rugged country. This mine and another opened by Youngstown Sheet and Tube near Logan, West Virginia, have the newest systems for safe operation and the prevention of air and water pollution. In all, twelve new mines, with an annual capacity of four million tons of coal, were opened on C&O/B&O."

"Giving C&O/B&O the sinew and muscle to help carry the nation's cargoes is a never-ending task. In the year, $118 million of new equipment was delivered, bringing the total to a half billion dollars in just the last four years."

* * * * *

"On March 20, Examiner Edward L. Boisseree recommended to the Interstate Commerce Commission that the merger of C&O and N&W be allowed, with certain conditions, and that the B&O, WM and other railroads be included. On November 6, after two days of oral presentations by all parties interested, pro and con, the eleven-member I.C.C. began its consideration of the proposal. The decision is expected early in the new year."

* * * * *

"Two directors on the C&O Board resigned at the end of the year. Barnum Colton, President of the National Bank of Washington, D.C., had been a director for sixteen years. Erwin H. Will, Chairman of Vepco, had been a director for eleven years and often disagreed with rail pricing policies."

"C. Vernon Cowen, Vice President of the Operating Group on C&O/B&O, was elected to the B&O Board in October to fill vacancy created by the death of Edmund Grimes."

* * * * *

In the fall of the year, I got a letter from Greg asking if I could use Rudnicki anywhere in my office. I studied the matter awhile and decided that since we had no engineer in the Fairmont office, he could be helpful in heading up that operation. Harold Beatty had been gone for several months and Rex Jones was acting head in his absence. I replied that I could use him in Fairmont and received a copy of DeVine's letter to Rudnicki advising him to transfer. Evidently the management interest in developing our own petroleum resources had greatly diminished and the program was being discontinued. Rudnicki reported to the Fairmont office on December 6 and continued to serve there for several years.

One of the calls I usually made whenever I was in Charleston, West Virginia was the Southern Land Company with offices in the Kanawha Valley Bank Building near the Daniel Boone Hotel. Turner Ratrie was in charge of the Company and was the son-in-law of the founder, C. C. Dickenson.

Southern owned several underdeveloped coal properties in the southern part of the state, one of which adjoined our Newberry Lands. During the summer on one of my visits, Turner told me that he was negotiating a lease with Cannelton Coal Company and said they wanted to include our reserves on the Newberry Lands in order to site a large coal mine. I told him that we would be glad to cooperate. Cannelton was noted as a progressive company with Paul Morton in charge. Southern had a reputation for making the most stringent lease agreements in the industry. I told Turner that he could advise them that we would probably be delighted to make a lease with them on the same terms as Southern.

This is exactly the way the deal worked out and construction of the mine started about three months later. In the meantime, we had a conference on the ground to plan the track layout for the tipple and the branch line of several miles to reach it. Bill Dixon was an assistant in the Engineering Department who came up through the ranks. He was a very capable siting engineer and always met with us at these meetings. In about 20 months, the mine was formally dedicated in a ceremony at the plant site. Greg flew in from Cleveland with several vice presidents and many other dignitaries attended. Clarence Pauley was the Chief Engineer of Southern and I was glad to congratulate both he and Turner on the success of our mutual project.

This mine could not have been sited at this location if the Raleigh County Commission had not been persuaded to change their lake site described in the review of the year, 1960, to Stephens Branch.

An incident occurred in the summer of the year which profoundly affected our department and must have caused great consternation throughout the company. I called the Cleveland office and asked to speak to Hewitt and his secretary said he wasn't in the office and she didn't know when he would be available. I then talked with Harold Lane, the Chief Clerk, and he gave me the distressing news that Hewitt had cleaned out his desk over the weekend and they weren't sure that he would be back. Subsequent discussion revealed that he was not going to return. Evidently he had decided that he could no longer endure the demands of his job and made the decision to leave. Later we learned that he had moved his residence to Phoenix, Arizona where a brother lived.

Due to the loss of our leader, we started reporting directly to DeVine again. George Sandmann was the number two man in our office and I started working more closely with him.

☆

1970 - WATKINS TAKES OVER AS PENN CENTRAL COLLAPSES

"Here is the box score for the year just ended, another in which C&O/B&O turned a profit for moving vast tonnages of the things America and the world need."

	1970	1969
Consolidated Net Income (millions)	$52.9	$52.5
Per C&O Share	$6.00	$6.05

* * * * *

	1970	1969
Operating Revenues (millions)		
Merchandise	$536	$533
Coal, Coke and Iron Ore	$348	$293
Passenger and other	$48	$48
Total Operating Revenues (millions)	$932	$874

* * * * *

"For a coal-hungry nation, greater tonnages were moved to electric plants, steel mills, export and other users. In all, 122 million tons were hauled on C&O/B&O, up ten million from the year before."

"With C&O/B&O cooperation, mine operators increased production, reactivated closed mines and opened new mines. Twenty new mines began operations in the year, with annual production at the rate of five million tons. Four smaller mines, producing a total of 362,000 tons yearly, closed."

"And the future of coal traffic looks excellent with continued heavy demand for this abundant energy source. Forty-six mines will open in 1971, with a capacity of eleven million tons yearly."

"To help do their part in moving a mountain of coal, C&O/B&O have kept a stream of new coal hopper cars joining their fleet. A total of 3,320 were added in 1970, and another 4,420 will roll into service during the next six months."

* * * * *

"Almost 2.5 million carloads of merchandise freight-everything from steel to paper cups and from chemicals to grain were moved. This was nearly the same as the year before and reflected the continuing slow pace of the nation's business along with work interruptions at major shippers and on the railroads."

"One hundred fifty-nine industries located on C&O/B&O in the year, and fifty-three on-line industries expanded their plants. This growth, representing a capital investment of $281 million, is expected to produce $21.5 million of additional revenues yearly for C&O/B&O. Plant closing cut annual revenues $1.6 million."

"To move the traffic, a total of 3,176 merchandise freight cars of all sizes, shapes and purposes were added in the year. Both merchandise and coal trains need power, and 200 diesel locomotives were received or on order."

* * * * *

"General Electric's vast Appliance Park East at Columbia, Maryland, is scheduled to begin the first phase of operations in the first quarter-eventual annual freight revenues to C&O/B&O are expected to be several millions of dollars. In the same revenue class is GM's immense car-assembly complex at Lordstown, Ohio, where a doubling of capacity will be completed early in the year."

* * * * *

Hays T. Watkins, Chairman of the Board and Chief Executive Officer

The Spring House (left) is a well-known symbol of The Greenbrier. (Right) The North Portico gives one of the most impressive views of the renowned resort.

"Equipment acquisitions in the year totalled $99 million. Another $23 million was spent for improvements to track and roadbed, buildings, bridges, freight terminals and electronic communications and traffic control systems. This brought capital expenditures to $122 million. For 1971, equipment orders so far total $98 million."

* * * * *

"January 19. Snow and ice smothers the Appalachians, keeping coal mines from work. The nation's coal output was slowed by weather for the rest of the month."

* * * * *

"At Penny, Kentucky, construction began of a 2.4-mile spur track to reach a new coal mine of the Bethlehem Steel Corporation. It will produce a million tons annually. Soon after, a five-mile spur track was started to a new mine at Dema, Kentucky, of the National Steel Company with production set at 1.5 million tons a year."

* * * * *

"Anheuser-Busch, Inc., broke ground for a $50-million complex on C&O near Williamsburg, Virginia. It should add $1 million a year to C&O/B&O merchandise revenues when complete."

* * * * *

"Island Creek Coal Company's new mine at Kohlsaat, West Virginia, expected to produce 1.4 million tons a year, loaded its first shipment into C&O trains."

* * * * *

"C&O loaded more coal in October than in any month since May, 1957. B&O's loadings were the best since August 1968."

* * * * *

"Reopening of the C&O-Norfolk and Western Railway merger case was ordered by the I.C.C. on June 16 for the 'limited purpose' of obtaining evidence on possible traffic diversions. C&O-N&W and Penn Central were directed to submit statements and exhibits on traffic diversions within ninety days. This was later extended to February 15, 1971, at Penn Central's request."

* * * * *

"Working Capital at the year end of $28 million was entirely sufficient to meet needs. It compared with $34 million at the previous year's end."

"The reduction was caused by the reclassification from current to non-current assets of $5.6 million in accounts receivable. All of this sum, related to the routine transportation activities of interconnecting railroads, is due to C&O/B&O from the Penn Central Transportation Company."

* * * * *

It had been anticipated by many in the Company that Hays Watkins would be designated as heir apparent early in 1970 so as to have plenty of time to prepare for the job as head man. During the latter part of that year, several stories started emanating from the Cleveland headquarters about indecision as to who would get the job. When Hays was finally appointed the new President, it was obvious that Greg would still wield a lot of influence as Chairman of the Executive Committee on the Board of Directors.

One of the practices in our department that always bothered me was that, every year or so, the head of our department would take a trip overseas to promote more business from potential customers to take their coal from the mines on our system. Several trips were made to Japan and a few to Europe. The only other people that went along, as far as I know, were a couple of his assistants in Cleveland or the President of the New River Company. Even though our office was the best source of information about the mines and quality of coals available on our system, I was never asked to go along. This made me feel that I was deliberately overlooked even though I knew that my knowledge and experience could have been very helpful in any discussion of the availability of coal and reserves and the very complicated characteristics regarding their quality and location.

Whenever we hired a new clerk in our office, we had to operate through the guidance of our Labor Relations Department to avoid trouble with the rules and restrictions required by the clerks' union which were obviously very demanding. One of the men in Labor Relations that I got to know quite well was Kenneth Stephenson. He was from Virginia and worked his way up in the engineering and operating departments in the southern coal fields. He was evidently promoted into labor relations due to his personality and ability to get along with people. He was very helpful when we needed him.

J. Hilary Kelley became Dean of the College of Minerals and Energy Resources at WVU soon after I was appointed to the Advisory Board. He was a brilliant engineer who had served President Kennedy as Science Advisor. We became good friends and he was able to attract and educate a record number of graduates for the coal mining industry. Gerald Spindler and Charles Holland had been fine leaders in the former School of Mines. John Schroeder was a classmate who served after Kelley. Larry Grayson and Royce Watts, his first assistant, are carrying on the great tradition.

THE C&O BOARD OF DIRECTORS

Cyrus S. Eaton .... Cleveland, Ohio
Cyrus S. Eaton, Jr. .... Cleveland, Ohio
Lawrence A. Kimpton .... Lakeside, Michigan
Gregory S. DeVine .... Cleveland, Ohio
Milton S. Eisenhower .... Baltimore, Maryland
Dr. Fay A. LeFevre .... Cleveland, Ohio
John E. Kusik .... Cleveland, Ohio
Roger H. Ferger .... Palm Beach, Florida
Hays T. Watkins, Jr. .... Cleveland, Ohio

THE B&O BOARD OF DIRECTORS

Cyrus S. Eaton	Cleveland, Ohio
John D. Biggers	Toledo, Ohio
John E. Motz	Baltimore, Maryland
Delos W. Rentzel	Washington, D.C.
Hays T. Watkins, Jr.	Cleveland, Ohio
Gregory S. DeVine	Cleveland, Ohio
C. Vernon Cowan	Baltimore, Maryland
James Parker Nolan	Washington, D.C.
Howard E. Simpson	Baltimore, Maryland
Milton S. Eisenhower	Baltimore, Maryland
John E. Kusik	Cleveland, Ohio
W. James Price	Baltimore, Maryland
Arthur W. Steudel	Cleveland, Ohio

THE WESTERN MARYLAND BOARD OF DIRECTORS

William P. Coliton	Baltimore, Maryland
C. Vernon Cowan	Baltimore, Maryland
Alonzo G. Decker, Jr.	Towson, Maryland
Gregory S. DeVine	Cleveland, Ohio
P. H. Glatfelter, III	Spring Grove, Pennsylvania
Robert B. Hobbs	Baltimore, Maryland
John E. Kusik	Cleveland, Ohio
George M. Leilech	Baltimore, Maryland
Norman C. Melvin	Baltimore, Maryland
Philip W. Ness	New York City, New York
William C. Purnell	Baltimore, Maryland
Hays T. Watkins, Jr.	Cleveland, Ohio
Kenneth E. Ekin	Baltimore, Maryland

WESTERN MARYLAND OFFICERS

Gregory S. DeVine	*Chairman of the Board*
W. P. Coliton	*President*
W. R. Downey	*Vice President-Sales*
G. M. Lelich	*Vice President-Operations*
N. C. Melvin	*Vice President and General Counsel*
H. L. Ward	*Secretary and Treasurer*
E. P. Holland	*Controller*
Y. R. Holman	*Director of Personnel*
R. R. Gunderson	*Chief Engineer*
J. C. Shewbridge	*Chief Mechanical Officer*
E. C. Whipp	*Director, Material Management*

LAWSON W. HAMILTON

VICTOR N. GREEN

HENRY G. DAVIS

STEPHEN B. ELKINS

1971 - N&W MERGER SIDETRACKED BUT REVENUES IMPROVED

On March 31, 1971, Gregory DeVine retired having reached the age of 62. He was succeeded by Hays Watkins who had come with the company in 1949, promoted to Treasurer in 1960 and elected to the Board of Directors as Vice President of the Administrative Group earlier in the year. Vernon Cowan was elected to Senior Vice President of C&O/B&O. DeVine retained his position as Chairman of the Board of B&O.

At the annual meeting on April 22, 1971, Hays presented his first report which is excerpted as follows:

"Nearly a million carloads of merchandise freight and coal and coke rolled over C&O/B&O in the first quarter, an encouraging start for the year."

"Helpful to the good showing were increased movements of automobiles and automotive parts. Coal tonnages moved on C&O/B&O were at a high level."

* * * * *

"Just as new factories grow along our tracks, new coal mines are opened up, too. Last year, twenty of them started up and produced five million tons. Eight more that will eventually produce eleven million tons of coal annually are being developed in West Virginia and Kentucky. To serve them, C&O/B&O are building thirty-six miles of new spur lines and eleven bridges."

"And in these first months of 1971, movements of coal continue at a high rate. Since the year began, we have moved twenty-six million tons. By year's end, C&O/B&O will have hauled an estimated 120 million tons for steel mills, electric power plants and other users at home and abroad. Of the total, eighteen million tons are expected to move this year overseas."

* * * * *

"Assuming, then, no major work interruptions during the rest of the year, we expect C&O/B&O to report improved earnings over last year's $6.09."

"Helping 1971's results-and, more significantly, those of future years-will be the elimination of our deficit on intercity passenger trains when the National Railroad Passenger Corporation, which has adopted Amtrak as the short form of its name, takes over on May 1."

"On April 19, I signed a contract on behalf of C&O/B&O to join Amtrak, at a cost of $9.5 million. That is equal to one-half C&O/B&O's 1969 passenger deficit of $19 million, figured on the Interstate Commerce Commission's fully-distributed-cost basis. We will pay that 'entrance fee' at the rate of $264,000 a month for three years."

* * * * *

"Since Amtrak will not take over commuter trains, B&O will continue to operate its Washington and Pittsburgh commuter services, which last year had a deficit of $1 million. C&O will continue to carry passengers and their automobiles, which last year contributed to income $1.4 million, on its Lake Michigan ferries for freight cars."

"When C&O/B&O joined Amtrak, C&O agreed to operate for it, under contract, those segments of our 'George Washington' train which now runs between Newport

News, Virginia, and Cincinnati, with a connection between Washington and Charlottsville. 'The George Washington' was the only C&O/B&O train selected for the new national system. All other C&O/B&O intercity passenger service will halt at the end of this month."

* * * * *

"And now we come to the termination of the C&O-N&W merger plan. With your proxy, since it was all we could manage in the time available, you received a copy of the brief news announcement. It stated that the N&W and C&O Boards would be asked to terminate the merger plan 'in view of the drastic changes in the Eastern railroad picture since the plan of merger was entered into in 1965.' N&W's Board voted to terminate the plan on March 23. I fully expect our Directors to do the same late this afternoon."

* * * * *

"In June, the starting bankruptcy of the Penn Central ended the threat of that giant system dominating the East. That threat had provided the principal incentive for the C&O-N&W merger. Meanwhile, because of changed stock market prices, the merger exchange ratio no longer offered a premium to the C&O shareowner."

* * * * *

"Happily, while the collapse of the N&W merger may have left both parties waiting at the church, it by no means finds your C&O/B&O left in a lurch. Our position is secure and our prospects are excellent. I would like to list for you, in conclusion, the basis of my optimism about the C&O/B&O."

"No. 1, the business outlook is most encouraging.

No. 2, C&O/B&O are in good physical shape.

No. 3, C&O/B&O's debt, while large, is manageable.

No. 4, the C&O/B&O are favored with major holdings of land for development.

No. 5, there are hopeful signs that national opinion leaders, the public and, most importantly, the Congress are increasingly aware of what must be done if the country's railroad industry is to provide the increased transportation service that will be required of it in the years ahead.

No. 6, the bedrock of my optimism about C&O/B&O is people. Although this is my last point, it really comes first. I speak of the 45,000 men and women who are my partners in C&O/B&O. They are dedicated to giving on-time, dependable transportation service. That is the only thing we have to sell, and it is what we will continue to strive for."

At the same meeting, Cyrus Eaton made a talk in which he praised the new leader:

"During the seventeen years in which it has been my privilege to serve as Chairman of the Board, it has been my good fortune and yours to have three unique occupants of the all-important office of President and Chief Executive. First there was the late Walter Tuohy, who probably had more friends than any other businessman in America. Then came Greg DeVine, whose hallmarks are warmth of personality and abundance of ability. I congratulate you and myself that, on Greg's retirement from presidential duties, we have in his place the remarkable individual who is the Chairman of this meeting today."

"Hays Watkins has spent all of his business life with our railroad. He is still a young man, in the prime of his life, with a wealth of experience and knowledge. He is the best do-it-now man I have ever known. He never puts off until tomorrow anything that can be done today. I predict a great future for him, and for the railroad industry of America, because of his genius."

On September 19, 1972, Hays wrote a letter to stockholders enclosing the third quarter dividend check. He also included a message about the new name for the company:

"Now C&O and B&O are Chessie System. And we have a brighter, more modern look. Our salesman said the popular, short-form system name for C&O and B&O would help them in their job of selling our services. Chessie, they said, 'symbolizes the care with which the patron's freight is handled.' Our safety people said a brighter design would increase visibility. Our operations and maintenance people said they were more than ready to place a new system designation and revised color scheme on our rolling stock. With that, our public relations people, who had produced the original study recommending the short system name, went to work to create the new color applications, the 'Ches-C' mark and the distinctive letter forms of the new signature."

* * * * *

"Firstly, it shows our readiness to welcome change, to adapt to the different in our technology, the way we do our job and the service we provide. Secondly, it confirms our dedication to safe, on-time rail transportation. While images, names and color scheme all are important, performance, as every Chessie System man and woman knows, is the most important ingredient of all in a good reputation."

* * * * *

"One of eighty-five new locomotives Chessie System is buying at a cost of $24 million is gold instead of the system's new color scheme of yellow and blue with vermillion stripes. The gold locomotive was the star of the 50th anniversary celebration held on September 8 at LaGrange, Illinois, by General Motors' Electro-Motive Division, which achieved a constructive revolution in rail motive power."

* * * * *

Consolidated Statements of Earnings

Income (millions)	1971	1970
Coal, coke and Iron Ore	$338	$369
Merchandise	627	574
Other Freight	51	52
Passenger Service	9	14
Dividends, etc.	15	38
Total	1,040	1,047
Expenses		
Payrolls	453	448
Materials	302	290
Allowance for Wear	66	67
Total	821	806

Earnings	28	53
Extraordinary Charges Less Taxes	87	6
Earnings (loss) for Year	(59)	47
Earnings Per Share		
Before extraordinary charge	$3.25	$6.09
After extraordinary charge	(6.83)	5.42

John R. Fields was a coal operator who came up in the ranks at Guyan Eagle Coal Company and later organized his own company, Clothier Coal Company on the C&O. When the Boone County Coal Corporation became idle because the owners were forced to leave the coal business, John was encouraged to take over by the C&O to preserve the traffic flow. He responded in a big way and operated the mine for many years.

A couple of coal operators on the B&O came to my acquaintance as leading citizens of northern West Virginia. Jim Compton was a very successful surface mine operator who invented a large boring machine to remove coal from a stripped coal seam. James LaRosa surface mined a large area of Pittsburgh seam coal near the top of Bridgeport hill which is now the site of shopping malls.

The new "Chessie System" name and the new color scheme on the locomotives was a big shot in the arm for morale improvement all over the Company. Hays Watkins has to be given credit for the adoption of these new ideas. The public also showed its appreciation for the new look in compliments and the quick adoption of the same color schemes on model trains appearing on the market.

Additions to coal development office during the seventies included Ronald C. Kopp, Jr., Guy E. Martin and Geoff Heath as coal field representatives. Karen Thrash was added to the clerk staff. Jennings D. Orwig was hired in the Fairmont office to serve as coal field representative. During this period there was a young man by the name of Pete Carpenter moving up noticeably in the operating department.

The main entrance to the Greenbrier is right on Route 60 just as you enter the small town of White Sulphur Springs. The main entrance is very impressive behind a large dazzling flower bed with massive columns. Several doormen are in green uniforms ready to unload your car. Since we are regular guests, they always review the incoming customer list and greet us by name.

Just inside the main entrance there is a pair of steps leading down into the cool and inviting shop floor where the main desk is located. The large black and white marble floor supports a steady stream of guests, new arrivals and departures.

Unit No. 4007 pulling the George Washington near St. Matthew, Kentucky, March 1968.
Photo courtesy of C&OHS.

Coal train led by G. E. U-30-B No. 8206 enters the yard at Clifton Forge, Virginia.
Photo courtesy of Mary Jane Rowe

The room clerk issues your room key to a green uniformed bellman who takes you and your luggage to your room. Every room is said to be decorated differently in the manager of Dorothy Draper who was a flamboyant designer from New York City after WWII. The rooms are large and have wallpaper with huge colorful designs of flowers.

Each room contains a booklet explaining all of the hotel services and the many sports facilities. My favorite facility is the large beautifully decorated indoor pool just off the shop floor. A large outdoor pool is operated in the summer season near the 'casino' which houses the golf shop and sumptuous dining area overlooking the golf courses. A new type high visibility bus makes continuous trips between the hotel and the 'casino.'

The shop floor beyond the hotel desk consists of a long hallway flanked by exotic shops on both sides. This is an area that provides the ladies with many hours of shopping pleasure while their husbands go for the more strenuous sports.

There is a fine shooting gallery up on the side of Kate Mountain across Route 60 from the hotel. This is a facility for skeet and trap shooting and a little further up the mountain is a fancy pavilion for private parties.

There is a beautiful trout stream, Howard's Creek, which passes through the grounds and provides great sport for anglers of all ages. The grounds between the hotel and casino contain a croquet court and many beautiful shade trees as well as cottages for resident employees and guests.

The main dining room of the hotel is a large beautifully decorated area which can accommodate several hundred guests at a time. A newly printed menu is available for breakfast and dinner with a piano and violin duo to entertain during the dinner hours. A popular spot for entertainment after dinner is a comfortable theater or a night club called the "Old White." A group of musicians called the "Populairs" play before and after dinner, the very danceable music preferred by the older generation. Underneath the Old White is located a supper club serving special hearty dinners and converts to a rock and roll club after dinner hours.

Features of the individual rooms include a writing desk, television set and small liquor cabinet containing drinks and snacks. The food and general atmosphere can only be described as wonderful.

After Frank Gaddy resigned, I obtained permission to hire another mining engineer. After making inquiries and advertising, I was able to hire Jose R. Bollar. He was a Cuban refugee who escaped from Cuba as a high school graduate along with a brother. He had resolutely pursued a mining engineering degree from the University of Idaho. He proved to be a very capable addition to our staff.

DIRECTORS

Cyrus S. Eaton	Cleveland, Ohio
Hays T. Watkins, Jr.	Cleveland, Ohio
Gregory S. DeVine	Cleveland, Ohio
John E. Kusik	Cleveland, Ohio
Cyrus S. Eaton, Jr.	Cleveland, Ohio
Milton S. Eisenhower	Baltimore, Maryland
Roger H. Ferger	Palm Beach, Florida
Lawrence A. Kimpton	Lakeside, Michigan
Dr. Fay A. LeFevre	Cleveland, Ohio

OFFICERS

Cyrus S. Eaton	*Chairman of the Board*
Hays T. Watkins	*President and Chief Executive Officer*
C. Vernon Cowan	*Senior Vice President*
John W. Hanifin	*Vice President-Executive Department*
Charles W. Campbell	*Vice President-Merchandise Freight Sales, C&O*
Owen Clarke	*Vice President-Labor Relations and Personnel*
William J. Eck	*Vice President-Purchases and Materials*
Kenneth H. Ekin	*Vice President-Law*
John T. Ford	*Vice President-Finance*
John P. Ganley	*Vice President-Taxes*
Charles J. Henry, Jr.	*Vice President-Commercial Development*
Edward M. Hudgins	*Vice President-Casualty Prevention*
Thomas A. Keefe	*Vice President-Merchandise Freight Sales, B&O*
Robert C. McGowan	*Vice President-Administration*
William L. Ollerhead	*Vice President-Industrial Development*
Kenneth T. Reed	*Vice President-Operations and Maintenance*
George A. Sandmann	*Vice President-Coal Traffic*
Howard Skidmore	*Vice President-Public Relations and Advertising*
C. R. Zarfoss	*Vice President-Merchandise Freight*
William R. Althans	*General Counsel*
Robert L. Hintz	*Comptroller*
T. Howard Keelor	*Corporate Secretary*
L. C. Roig, Jr.	*Treasurer*
E. Truman Wright	*Vice President and Managing Director The Greenbrier*

★

1972 - AUSTERITY STARTS PAYING OFF

The financial review for the year shows that the Chessie System is making steady progress:

Summary of Earnings (Millions)	1972	1971
Income		
Merchandise Revenues	$642	$627
Coal, Coke and Iron Ore	331	338
Other Freight & Pass Revenues	52	60
Total	1025	1025
Other Income - Net	21	18
Total Income	1047	1043
Expenses:		
Payrolls	447	454
Materials, Supplies, etc.	273	301
Depreciation & Retirement Losses	67	66
Property & Other Taxes	78	76
Equip & Joint Facility Rentals	56	59
Interest on Long Term Debt	55	57
Total Expenses	975	1012
Earnings Before Income Taxes, etc.	71	31
Income Taxes	15	(2)
Extraordinary Items	(4)	(88)
Earnings for the Year	53	(55)
Earnings (Loss) Per Share		
Before Extraordinary Items	6.49	3.77
Earnings (Loss) for the Year	6.08	(6.31)

The figures shown on the table above have been slightly altered from those originally published in the flash annual report and reflect deferred Federal Income Taxes.

It is obvious from these figures that the improved earnings resulted from a reduction in expenses. Income was only $4 million more than in 1971 but expenses were reduced $37 million. A very severe austerity program had been initiated in 1971 when it became obvious that a change in accounting procedures would result in huge loss in earnings which in turn would force the Board to reduce dividends from $4 to $3.

At the annual meeting held in Cleveland in April of 1973, Hays Watkins made a lengthy talk and slide presentation about the business prospects of the Company.

"Predicting the weather obviously is as uncertain as predicting the economic climate. The experts who issue business forecasts predicted that the pace of sales and production would speed up early in the year."

"Those market-place meteorologists turned out to be no more reliable than the TV weathermen at times. The nation's economy did not pick up in the first quarter. As a result, C&O/B&O carloads of coal and merchandise freight were off by seven percent, and revenues were down by $12 million from the same period last year."

"Yet in spite of the squeeze of higher costs, C&O/B&O managed to report record earnings for the year's first three months. Net income was $11.4 million, equal to $1.31 a C&O common share, up substantially over the $8.8 million, or $1.02 a share, in last year's first quarter. It was the highest net for any first three months in the ten of years C&O/B&O affiliation."

"The good results were achieved only through a concerted effort to hold the line on expenses."

"We hope now that the long-awaited business recovery is finally on its way. Our C&O/B&O market specialists, who follow closely the fortunes of the various industries which we serve, see no dramatic over-night boom. Rather the outlook is for a slowly improving economy resulting in direct benefits for C&O/B&O business for the rest of the year and beyond."

* * * * *

"Earlier this year the Chairman of United States Steel and the Chairman of Bethlehem Steel both said that the steel industry's prospects for a good 1972 hinged on its performance in March. Well, steel production in March did move upward steadily, and it has continued to do so this month. In fact, the output from all mills in April's first two weeks was the highest in ten months."

* * * * *

"A herald of longer range business growth for C&O/B&O is the development of new industries along our lines. In the year's first quarter, forty-nine companies picked sites served by our railroads for new plants or the expansion of existing facilities. To C&O/B&O's future merchandise revenues, these will add an estimated $3.4 million a year."

"Right here in Ohio, which generated one-fourth of the industrial development along C&O/B&O last year, a corn products refinery is being built near Dayton by Cargill that will eventually contribute, we expect, 11,000 carloads of merchandise freight to C&O/B&O annually. Near Akron, a liquid sugar refinery that will generate over a quarter-million dollars a year in freight revenues is planned by Amstar. Just a short way south of here, General Motors is expanding substantially its drive shaft and transmission plant. And at Fostoria, west of here, General Mills has just centered all of its midwest distribution operations at the huge food consolidation complex served by C&O/B&O."

* * * * *

"Just as pickup in the steel, automobile, chemical and other industries will help our merchandise freight, so will it be beneficial to our coal traffic. In the first quarter, revenues from coal movements were down four percent from last year's same period, when a boom in coal demand was being felt."

"A base of our optimism about coal is the rich and plentiful supply of steel-making coal in our territory. Of the twenty-two mines that will be started up this year on our system, eight of them will ship their product to steel mills."

* * * * *

"800 new open-top gondola cars are being built at the company's shops at Raceland, Kentucky. Half of them are specially equipped to transport hot ingots from blast furnaces to rolling mills over distances as great as 400 miles."

"Seventy-five 'Canstock' boxcars designed by C&O/B&O are in service carrying coils of thin steel used for making cans."

"New 'Vert-A-Pak' cars are transporting new automobiles over C&O/B&O under complete cover. Also, 'Stak-Pak' cars, which carry autos stacked three high in containers, are being tested."

"C&O/B&O's vast automobile distribution terminal at Jessup, Maryland, is being enlarged to provide room for 400 more autos."

"TBS terminals, originally designed for handling liquid and dry bulk chemicals, are now being used for shipments of plastics, grain, corn syrup, minerals, and - if a proposed plan works out - will also handle Scotch whiskey."

"Computers are being used to monitor day-to-day performance of freight trains and the individual cars in them, enabling C&O/B&O to spot quickly and eliminate hitches that can interfere with good service."

"Besides $34 million being spent for 125 new locomotives and $66 million for freight cars, an additional $20 million will be expended on additions and improvements to physical facilities. This makes a total of $120 million in capital expenditures in 1972."

* * * * *

"In just one year, therefore, C&O/B&O are spending a grand total of $400 million to buy new tools and to preserve the tools we already own."

* * * * *

"Threatening selective railroad strikes is the Sheet Metal Workers Union. This union of less than 6,000 members out of an industry total of half a million was the only one to reject the wage increase already accepted by all other unions. The sheet metal union not only wants more money than the 42 percent wage increase settlement, but seeks repeal of a work-rule change imposed as a result of Congressional action."

"A second strike threat is against the Penn Central by the United Transportation Union, protesting that bankrupt railroad's plan to eliminate 6,000 positions by phasing them out as the present holders retire or move to other jobs. A shut-down of Penn Central would affect C&O/B&O and other lines, interfering with the interchange of freight movements."

* * * * *

"In January, the railroads applied to the Interstate Commerce Commission for authority to place a surcharge of 2.5 percent on most freight shipments. This was granted as a temporary measure and took effect on February 5."

"In March, the railroads asked to raise rates selectively by an average of 4.5 percent, effective April 1, including the emergency surcharge already in effect. The I.C.C. has ordered hearings held with no increase effective before May 1. And just this week, the Price Commission asked the ICC to delay any freight rate increase for another thirty days."

"If the Commission acts affirmatively, C&O/B&O's 1972 operating revenues could be increased by about $30 million."

"I might add that the rail industry's major competition, the trucking industry, is also seeking increases in the rates it charges, so we won't be pricing ourselves out of the ball game."

During the year, two very capable veteran officers retired and it was obvious that their wisdom and experience would be missed. John E. Kusik retired as head of the Finance Department and was replaced by John T. Ford who had been one of his very capable assistants. During the ensuing years of my career I got to know and admire "Jack" very much for good judgement and personal friendship. He was a man small in height, who had a good upbeat personality and was his own man.

The other retiree was Vernon Cowan who I never got to know personally but was aware that he had many friends and admirers among the people who worked for the Company. He was succeeded by Kenneth T. Reed who had been a member of the operations department on the B&O for many years.

In November, my second wife Claire filed for divorce. It was both a shock and relief to me because our personal goals were diverging.

When I left home to fly to Pittsburgh to attend the annual Coal Show, my second marriage was in a shambles but I had just been advised to take over the Coal Development office of the Western Maryland Railroad in Thomas, West Virginia which boosted my spirits somewhat. I was looking forward to this new challenge and planned to go there during the following week.

While talking to some friends in the lobby of the William Penn Hotel, I was approached by a rough-hewn individual whose features indicated his Italian heritage who introduced himself as Carman Del Signore and invited me to join he and his brother at the bar for a drink. The brother was Carl, the President of Buffalo Coal Company located in the small town of Bayard, West Virginia on the Western Maryland main line. He was roly-poly individual with a bald pate and a winning smile. Both men were obviously well lubricated and feeling no pain. They had somehow found out that I was taking over the development office in their area and were losing no time in putting the bite on me.

I will have to say that this was the beginning of one of the most pleasurable and rewarding friendships of my entire life. They insisted that I come to their office at the first opportunity and they would show me around the field. I assured them that I would contact them as soon as I could and take them up on the deal.

While still at the Coal Show I ran into Al Salvato who was the traffic representative of the WM who had previously worked in the coal field and he offered to meet me on the ground the following week to meet the office staff in Thomas. At the office, I met Paul Dumire who was a jack-leg engineer but proved to be one of the best staff members I ever worked with. He was a small man with rough-hewn features who spoke with a grovelly voice and we hit it off immediately. His secretary, Mary Baltimore, was very efficient and most helpful.

When I got back in the field to visit with the Del Signores, Carman insisted on giving me a tour of the field. He had a fancy jeep station wagon with only one speed, 55 MPH which wasn't too bad on the open highway, but a little upsetting while going through towns or dirt roads and surface mines. He had the trip well planned so that we had a drink at a friend's trailer at 10:00 a.m., another at lunch and one at 3:00 p.m. in a bar. He kept up a running commentary as we traveled and it was very obvious that he was a very smart individual and knew a great deal about the coal business. When we stopped at his brother Frank's surface mine, a man came running up to our jeep shouting - "Get that SOB off of my property." It tuned out that Frank thought I was the local mine inspector.

After a few weeks on the job and conferring with Paul Dumire, I became convinced that the Del Signores were excellent miners as evidenced by their production records. I then directed Paul to embark on a program to lease to them every available reserve that they agreed to develop. We also had two other small operators who were good miners and we tried to help them all we could.

Al Salvato met me in Thomas, West Virginia to meet the local office staff. Paul Dumire was a man in his early fifties who had come up through the ranks and had worked for the WM for a long time. His secretary was a widow, Mary Baltimore, and both were very competent individuals. Their office was a very old brick building built by Davis Coal and Coke Company whose lands and mines were purchased by the WM railroad. Henry Gassaway Davis, who established the Company, was a West Virginia coal pioneer and could have used the same office. He was one of the founders of the WM.

Richard Nixon was re-elected, but the Watergate scandal was closing in on him.

George Sandmann and I had a long mutually successful association from the time in 1960 when I took over the coal development office. He then was the first assistant to Greg DeVine the Vice President of our coal department. When I couldn't reach DeVine, I always relayed my message through Sandmann and as Senior Assistant, he actually called a lot of the shots in the Boss' office.

When I took over the Development office of the B&O in 1966, the land companies were managed under a subsidiary company called Mid Allegheny Corporation. It was headed up by an elderly engineer by the name of S. Dunlap Brady in Summersville, West Virginia which was much closer to their largest properties. This operation was not included when I took over the B&O coal development office and Dunlap continued to report to our Cleveland office. George and I often discussed the fact that we should make arrangements to have Dunlap report to me to reduce the case load on the Cleveland office. In the meantime, Dunlap had obtained permission from Cleveland to hire an assistant engineer who would naturally succeed him since he was already in his late sixties. The man selected was David Morrison the operating manager of two mines located on B&O lands.

On one of my visits to the Cleveland office that year, George advised me that Dunlap would also be present. I assumed that this must be the time that I would be designated to take over the management of the Summersville office. However, it soon became clear to me that day during our meetings that no change was going to take place and George later confirmed to me that Dunlap would continue to report directly to him in Cleveland.

I didn't realize the fact until sometime later, but George really did me a favor in this instance. It so happened that Mid-Allegheny ran into great difficulty and expense after negotiating a lease for a new mine to develop their largest remaining coal reserve. The surface over this reserve was part of the Monogahela National Forest and the environmentalists got into the act. After months and years of frustration and negotiations, a trade of reserves was consummated but the original mining plan was wiped out.

Henry Gassaway Davis, along with his son-in-law Steven B. Elkins, founded several railroads which later became part of the Western Maryland. He served two terms as Senator from West Virginia and was nominated by both parties for his second term. He also formed the Davis Coal and Coke Company which he sold to the WM in 1902. The town of Davis was named for him.

☆

The classic Victorian style B&O passenger station in Oakland, Maryland

The former Davis Coal and Coke Mine office building in the foreground
where the WM coal development office was located.

1973 - COLLINSON PROMOTED AS EARNINGS CLIMB

Income increased more than expenses to make a much better financial picture for the year:

	1973	1972
Summary of earnings (Millions)		
Income		
Merchandise Revenue	$710	$642
Coal, Coke and Iron Ore Revenues	349	331
other Frt & Pass Revenues	65	52
Total Railway Op. Revenues	1124	1025
Other Income, Net	32	21
Total Income	1156	1047
Expenses:		
Payrolls	486	447
Materials, Supplies & Others	302	273
Depreciation & Net Retirement	67	67
Property & Other Taxes	91	78
Equipment & Rentals	65	56
Interest on Long Term Debt	53	55
Total Expenses	1062	975
Earnings Before Income Taxes	94	71
Income Taxes (Current & Deferred)	35	15
Extraordinary Items	8	(4)
Earnings for the Year	67	53
Earnings (Loss) Per Share		
Before Extraordinary Items	$6.78	$6.49
Earnings for Year	7.71	6.08

From the foregoing table, it can be seen that the merchandise revenues improved more than coal. To those of us in the coal department, it was obvious that coal production was on the upswing and prices were improving.

In the coal development office, we were getting many requests to build a loading track to put in a coal loading facility. This was more prevalent in the Kentucky district when non-union truck mines were proliferating at a rapid rate. Kenneth Reed was promoted to Senior Vice President to make way for a new man in operations.

John Collinson was elevated to the Position of Vice President in charge of Operations and Maintenance. He had been in the engineering department of the B&O before the merger and later promoted to Chief Engineer of the Chessie System. When he came to Huntington to take over the engineering department on the second floor of the OH building, I went to visit him to make his acquaintance and offer any assistance I could to help him get acquainted with his new position. I found him to be singularly uncommunicative. After I realized that I was the only one talking, I retreated to my own office.

In another personnel development, George Sandmann was promoted to Senior Vice President in charge of the coal department and Henshaw was promoted to Vice President. This promotion made no change in the way our office operated and we continued to report to Sandmann.

The other news in personnel for the year was the elevation of John Hanifin, to the Presidency of the Chessie System although Watkins retained the designation of CEO. This action preceded and triggered the promotion of Sandmann and Henshaw who were personal friends of Hanifin, and all were territories of Greg DeVine.

A major problem was brewing for the system when several wrecks made headlines during the year. Most of them were occurring on branch lines which served coal operations. While conflicting reasons were given for the cause, it was my own opinion at the time that track maintenance efforts should be expanded.

My divorce decree became final in January and sometime thereafter I contacted a divorcee, Velma Neale, who had been one of my first wife's best friends. Although she was several years younger, we seemed to have much in common and after dating steadily for several months, were married November 16. She had one son, Stephen, who was married in the same year before us. He obtained employment teaching in the local school system and his wife, Letha, a graduate nurse, started to work for a local obstetrician. My own son, Brent, had married a girl while stationed at an army ammunition depot in the mountains of Nevada. He had brought his bride back to Huntington and enrolled in Marshall University. She obtained employment in a local legal office.

The annual report indicated that Cyrus Eaton had been named Chairman Emeritus. He and his son were still listed on the Board of Directors.

The coal business was much better than average when an event on the other side of the world triggered a boom in coal that exceeded anything I had seen previously. The OPEC nations made an agreement to severely restrict the export of petroleum. This action soon was reflected in growing apprehension among our citizens who feared a shortage of gasoline to fuel their cars. It wasn't too long that reduction in gasoline supplies did develop and long lines started to materialize at gas stations.

Although this condition only lasted a couple of months, it was enough to convince most people in the coal industry that alternate fuels would have to be used to supplement gasoline fuels. Major oil companies started to acquire coal reserves and operations. Utilities started to look more to coal as people became aware of the hazards of atomic power plants. Foreign

H7b No. 758 with the westbound freight sets out at Thomas, West Virginia. The former passenger station and power plant appear in the background. Bert Pennypacker.

countries started to look to superior quality U. S. Metallurgical coal to blend with their own inferior supplies. The Japanese started to import large quantities of mild-volatile coal out of the B&O Somerset field to mix for their steel mills. All of these factors caused a steady rise in the price of coal which translated into greater profits for the operators.

Soon after the announcement of promotion to Vice President for Sandmann, I learned that another promotion of a quite different sort had taken place. Ross Conlin was notified that he was to receive a new assignment as assistant to our government liaison representative in Washington, D.C. The man in charge there was John Snow, a young lawyer who had worked several jobs in the federal bureaucracy. He had been hired a couple of years before to lobby in our behalf. Although Ross' outgoing personality and railroad experience made him a good candidate for this type of assignment, it was apparent to me that this was not the main reason. I had often reflected in the past that Ross was the ideal candidate for Vice President of coal traffic and hoped that by some miracle his suitability would be recognized.

Ross never complained or expressed any disappointment to me. We had dinner together a couple of times over the next year, but the great expectations had been drained out of our friendship.

Another sad note was introduced into my life when Dr. Lawall was found dead of a heart attack in his apartment. I knew that I had lost a great friend and mentor that few men are privileged to have in a lifetime. His wife had died in 1970 and I had watched him slowly deteriorate in health and spirit after that event. During the company austerity programs in 1967, Greg ordered me to advise him that his monthly consulting fee would be terminated. This was an onerous chore for me and I knew that he would be sorely disappointed. Walter Tuohy had provided him with a post retirement income to set up his own small office with a secretary in the WV Building. Walter probably reasoned that Dr. Lawall would be helpful to me in an advisory capacity. I made it a point to call on him as frequently as I could. Sometimes I had a problem which I sought his advice. He always listened earnestly, but obviously avoided the slightest hint of advice in every case. This was very frustrating for me at the time, but in retrospect, I can see that he was giving me the opportunity to stand on my own two feet and give me the confidence to handle my own responsibilities.

On all my reports to Cleveland, I always ended with a recommendation regarding the subject at hand. I would guess in retrospect that my advice was essentially accepted at least ninety percent of the time. One failure occurred at a time in early September where one of my field men brought the news that two mines owned by Zapata were for sale at half price. The mines were owned by a family trust and a court had ordered that the two mines had to be divested. It was right in the middle of the coal boom and I immediately obtained as much information in a few days and realized that this was a terrific bargain. I wrote the matter up to Cleveland and was instructed to bring all my information to the meeting of the Rodeo at the Greenbrier for further consultation.

Late Thursday afternoon at the beginning of the Rodeo, George Sandmann advised me to meet in his room at 4:00 p.m. When I got there Greg DeVine, John Hanifin, Sandmann, Henshaw and Alex Hunt were there to hear my presentation. After I expounded for about 20 minutes, I told them these mines could be acquired for about 16 million dollars. I recall I waited for a decision. After only a brief discussion Hanifin expressed the verdict that the expenditure was too large to consider and the meeting was over.

Of course, I was disappointed by this turn down and later realized that I had not proceeded in the most promising manner. I should have discussed the proposition first hand with Alex Hunt to get his cooperation before the meeting. His endorsement could have made the difference. About a year later, Hanifin admitted in another meeting that we should have taken the Zapata offer.

☆

1974 - MASSIVE UPSURGE IN MINE DEVELOPMENT

A stock split was made on November 27, 1974 two shares for one, which is reflected in the following summary of earnings in the flash annual report. The report also reflects an increase in the dividend for the new shares from $1.75 to $1.875:

	1974	1973
Summary of earnings (Millions)		
Income		
Merchandise Revenue	$768	$710
Coal, Coke and Iron Ore Revenues	476	349
other Frt & Pass Revenues	77	65
Total Railway Op. Revenues	1321	1124
Other Income, Net	41	32
Total Income	1362	1156
Expenses:		
Payrolls	505	486
Materials, Supplies & Others	373	302
Depreciation & Net Retirement	79	67
Total Railway Op. Expenses	957	854
Property & Other Taxes	115	91
Equipment & Joint Facility Rentals	85	65
Interest on Long Term Debt	52	53
Total Expenses	1209	1062
Earnings Before Income Taxes and Extraordinary Items	154	94
Income Taxes - Current & Deferred	63	35
Earnings Before Extraordinary Items	90	59
Extraordinary Items	5	8
Earnings (Loss) for the Year	95	67
Primary		
Earnings Before Extraordinary Items	$5.06	$3.39
Earnings (Loss) for the Year	$5.33	$3.85

Hays Watkins and John Hanifin issues a joint statement in the annual report on February 18, 1975 which is excerpted Here:

"The Chessie System - the C&O, B&O and Western Maryland - has long been the nation's leading transporter of bituminous coal. In 1974, our railroads moved 83 million tons from mines on their lines. Counting coal moved in cooperation with connecting railroads, we handled a total of 100 million tons."

"Reflecting the massive upsurge in mine development and expansion in our railroads' territory, a 50 percent increase in coal production along our lines is projected over the next five years. The improved levels of coal traffic so far this year on our railroads bear out the projection. Coal originations in 1975 should be even more than 10 percent above last year's."

"As the estimate of growth of Chessie coal traffic became increasingly firm in recent months, we reviewed our estimates of capital needs. In June, it was announced that expenditures totaling $300 million would be required, mainly for coal-carrying cars and motive power. It is now evident that increased outlays will be necessary to meet the enlarged traffic, and might total a billion dollars in five year's time."

* * * * *

"Merchandise freight traffic constitutes the greater part of Chessie's business, contributing nearly 60 percent of railway operating revenues. On the basis of just merchandise freight tons transported, Chessie lines rank fourth among all railroads."

"Merchandise traffic volume so far in 1975, tied as it is to the national economy, is running below the levels of 1974's first quarter."

* * * * *

"Throughout 1974, Chessie roadways, structures, equipment and communications and signal systems benefited from sustained programmed maintenance. This maintenance continued even during the coal miners' strike, when the source of nearly 40 percent of Chessie revenues was affected."

"This month, the 5,000th mile of continuous welded rail was installed on our system. Welded rail - long-lasting, easier-riding and requiring less maintenance cost - now constitutes more than half of our system's key trackage. Only one other railroad, the Santa Fe, has reached the 5,000-mile mark, and that by only a few steps ahead of Chessie."

"Maintenance of equipment programs were also pursued in 1974, and even more would have been done if additional supplies of steel and parts had been available. At the end of 1974, the ratio of Chessie coal cars awaiting repairs was only 1.7 percent of the total of such cars, and the percentage of all freight cars awaiting repairs was well under the average for the railroad industry."

* * * * *

"Added payroll costs will result from agreements to be concluded this year with the railroad unions. Higher costs of materials, fuels and purchased services added $50 million to 1974 expenses, and prices are sure to keep going up."

* * * * *

"This letter has dealt with our railroads, which contribute almost all of Chessie's revenues and are, therefore, mainly responsible for its earning power. However, 1974 was a productive and profitable year for the other members of the Chessie family as well."

"These are Chessie Resources, Inc., our forest resources, real estate, investment, and oil and gas exploration subsidiary; Western Pocahontas Corporation, the main unit of our coal-land development group, and the White Sulphur Springs Company, which owns and operates the magnificent Greenbrier resort complex in West Virginia."

A personnel change which was most important to our office was the retirement of George Sandmann. Henshaw succeeded him as head of the coal department.

John T. Ford was now listed as Senior Vice President and we were advised that he would coordinate both the coal and merchandise traffic departments.

For the first time, Robert L. Hintz was listed as Vice President of Corporate Services.

Early in the year a study of coal reserves in a large area south of Beckley was made to determine if a Beckley coal bed reserve could be proven for a new mine. We conferred with Tom Howard who was in charge of the Beaver Land Company which had some ownership in the area. He furnished additional information which was very encouraging and as a result we recommended to Cleveland that a drilling program be started and it was approved. I directed Jim Lilly to supervise the effort and he went about getting bids from drilling contractors.

John Hanifin had been a friend of mine ever since we worked together on my testimony in the B&O control case. On several occasions when I was in Cleveland, I would go to his office in the law department to visit.

The drilling program got underway with two drilling contractors in the fall and initial results were very encouraging. Each time I came to Cleveland to report on progress, John came to sit in on our meetings. We were about halfway through the program on January 24 when Jim Lilly and I went to the Cleveland office to report on the drilling as well as other current matters. George Sandmann advised us after we reported to the twenty-first floor that Hanifin wanted to see us in his office on the 36th floor.

We proceeded to the thirty-second floor where we transferred to a small private elevator which served the upper executive office. John's office was very plush and had a sitting area in front of a fireplace. His secretary ushered us into his presence and retired. He hardly greeted us before embarking on a tirade that we were dragging our feet on the drilling program and were putting him in an untenable position because the results would not be available to present at the annual meeting in February.

I was completely stunned by this "dressing down" because I had never even heard that the results were to be presented to the stockholder's meeting.

Finally we were dismissed with the admonition that we had better have the drilling results very soon. When we got back to Sandmann's office, he seemed to know what had happened. The only thing I could think of was that maybe we could accelerate the drilling with extra monetary incentiveness. I mentioned this to George who agreed to the approach by quoting the cliche, "they say money talks." We went back in the field and I offered each one of the six drilling crews an extra three hundred dollars if they completed the project in the allotted time. I have never witnessed a more magical transformation of workers before or since. The deadline was met with a few days to spare.

About three months later, I was requested to make one of my regular trips to Cleveland and was advised that a company plane would be at the Huntington airport and I could come up with it. When I arrived at the airport with my wife and her son, Steve, we learned that Hanifin had delivered the commencement address at Marshall University that day and was the one using the plane. The plane was a brand new Grumman executive plane with two huge jet engines. I felt like I was being pressed into the back of the seat when we blasted off the runway. Our flight to Cleveland only took a half hour.

I spent the night in the Cleveland Hotel that was my custom and hardly slept at all because I was carrying in my brief case a large map prepared in our Western Maryland office that demonstrated the potential for seven new coal mines in the vicinity of the North Branch of the Potomac River. This was in addition to two mines on the western side of the area already under development. The coal industry was in the process of the biggest expansion I had witnessed in my career and I thought this information so promising to the future of our company, that I would probably receive recognition for this promising development. No such luck.

By mid-year, the coal boom was in full swing. Our office was being besieged by companies and individuals trying to initiate or increase coal production. A typical example would be a trio of individuals who came into our office with a map showing boundaries of a coal reserve that could be leased and requesting that the railroad build a siding at the nearest point where the coal could be loaded into railcars. After we explained to them that a new siding would require almost a year to design and construct, their dreams of immediate wealth would vanish and their spirits became depressed. We also had to explain that we had to consider all applications in order of receipt and we already had a lot on file. In one area alone, we had a total of thirty requests for new sidings in Eastern Kentucky. Our management knew that all coal booms come to an end and that most of these operations were ill-conceived and underfunded so that most of them would never materialize. The whole process put a severe strain on the operation of our office. Along with all the "fly-by-nights," we had many legitimate applications which were handled and put into effect. We averaged installing new capacity of several million tons a year over the remainder of the decade. Unfortunately for the industry and the railroad, most of this capacity turned out to be more than the market could absorb and much money and effort was eventually wasted.

During this period, I was spending a lot of time getting information and making new leases on the Western Maryland properties. My good friend, Carl Del Signore was increasing his production every month and accumu-

The modern preparation plant of Buffalo Coal Company served by the WMRR. Separate piles of raw coal in top left. Clean storage on right side.
Coal being loaded into coal car in lower front area.

ONE CHARLES CENTER, Chessie's headquarters in Baltimore, Maryland

lating great wealth in the process. He was smart enough to develop mines with reserves that could be blended for different markets. His brother, Frank was anxious to retire and agreed to sell his mine to Carl. It was a wise move because part of the production was very low in sulphur and this could be blended with other coals for the net market overseas. The main problem was two-fold. Some of the reserves were too high in ash content and others in sulphur content. By maintaining stockpiles of several different coals at his tipples, Carl could blend them to make an acceptable product.

Carl had already lost one brother who was injured in WWII and died before I came on the scene. His brother, Carman who had initiated me in my first trip in his field, died in May of the year. These three brothers had been together in the coal business for several years and their passing was a severe loss to Carl.

A good part of Carl's success came from having a good staff. Don Cussins was his chief assistant with a good helper by the name of Denny Dolachek. All of his superintendents were good administrators. His engineer, Steve Shaeffer and his pilot, “Buzz” Gosnell handled their jobs with excellent results.

It dawned on me the past year that Paul Dumire was making a good contribution to my office in Thomas and was covered up with work. I started to work on the idea of obtaining an assistant for him. I broached the idea to George Sandmann and got his okay to proceed. Knowing that young mining engineers are hard to entice by previous experience, I decided to try to obtain a graduate in Geology from WVU. The head of the department suggested a young man named Jack Britton who had worked as a clerk for the CIA in Washington and Rome and came back to school to get his degree. After I decided to hire him, the personnel staff in Cleveland requested that I bring him in for an interview. Evidently they were concerned about his government background, but they decided to hire him. He proved to be a great help to Dumire and was most helpful to me when Paul came down with cancer in 1979.

Word came down to me in the middle of the year that our forestry personnel were to be transferred over to Chessie Resources. After all the work I had done to get a good forestry program in force, it was out of our office without any consultation where it had been logically and efficiently operating for several years.

In August, I received a phone call from Ted Wentworth in Cincinnati, Ohio inquiring about the availability of very low cost coal on the Chessie System. I told him that I couldn't give him a quick answer, but would think about it and call him back. There is one large area on the Chessie System where there are large reserves of poor quality coal that has no market. This reserve can be recovered at very low cost.

I obtained several samples and sent them to Cincinnati. The word came back that these samples indicated that the product could be ideally used to produce methanol.

Wentworth was working on a project to produce methanol to be used as a fuel for the boilers of large electric utilities along the east coast to replace fuel oil.

Carl Del Signore agreed to work with Wentworth and a report was prepared which concluded that the idea was economically viable. Several utilities were contacted and after showing a lot of interest finally gave up.

I tried to get the management of Chessie System interested in the idea to fuel their locomotives, but met with no encouragement at all.

I have kept in touch with Wentworth even after I retired and learned that methanol is a better fuel for automobiles than gasoline.

It is my firm conviction that the major oil producers will eventually come around to the manufacture and sale of methanol from coal to supplement our declining reserves of petroleum.

CHESSIE SYSTEM, INC. EXECUTIVE OFFICERS

Hays T. Watkins, Jr. .... *Chairman of the Board and Chief Executive Officer*

John W. Hanifin .... *President*

John T. Ford .... *Senior Vice President*

Owen Clarke .... *Vice President-Law*

John P. Ganley .... *Vice President-Taxes*

Norman G. Halpern .... *Vice President-Labor Relations, Personnel and Organizational Planning*

Robert L. Hintz .... *Vice President-Corporate Services*

T. Howard Keelor .... *Vice President and Secretary*

Howard Skidmore .... *Vice President-Public Relations and Advertising*

Doyle S. Morris .... *Assistant Vice President and General Counsel*

L. C. Roig, Jr. .... *Assistant Vice President and Treasurer*

Bruce G. Lawler .... *Assistant Vice President and Comptroller*

CHESSIE SYSTEM EXECUTIVE OFFICERS

Hays T. Watkins .... *Chairman of the Board, C&O; Vice Chairman, B&O*
John W. Hanifin .... *President and Chief Executive Officer, C&O, B&O*
John T. Ford .... *Senior Vice President*
Roger F. Brown .... *Vice President-Executive Office*
Owen Clarke .... *Vice President-Law*
John T. Collinson .... *Vice President-Operations and Maintenance*
William P. Coliton .... *President-WM*
John P. Ganley .... *Vice President-Taxes*
Norman G. Halpern .... *Vice President-Labor Relations, Personnel and Organization Planning*
Charles J. Henry, Jr. .... *Vice President-Merchandise Pricing*
H. Preston Henshaw .... *Vice President-Coal Traffic*
Robert L. Hintz .... *Vice President-Corporate Services*
Edward M. Higgins .... *Vice President-Casualty Prevention*
T. Howard Keelor .... *Vice President and Secretary*
James B. McCahey, Jr. ... *Vice President-Merchandise Traffic, C&O, B&O*
Robert C. McGowan .... *Vice President-Administration*
Howard Skidmore .... *Vice President-Public Relations and Advertising*
Doyle S. Morris .... *Assistant Vice President and General Counsel*
L. C. Roig, Jr. .... *Assistant Vice President and Secretary*
Bruce G. Lawler .... *Assistant Vice President and Comptroller*

☆

1975 - LOOKING FOR AN UPTURN

The flash annual report published on January 2, 1976 opened with two special comments:

"Your company's revenues and earnings for 1975 were down slightly from the all-time highs of 1974. However, they were better than in any other previous year."

"Encouraging signs are seen of a recovery of the nation's economy. Business should continue to improve in the months ahead. The Chessie System railroads, strong both physically and financially, are in an excellent position to benefit from the upturn."

"The financial results for 1975's full year are presented in the Summary Statement of Earnings."

	Twelve Months 1975	1974
INCOME		
Merchandise Revenues	$711	$768
Coal, Coke and Iron Ore Revenues	501	476
Other Freight and Passenger Services	77	77
Total Railway Operating Revenues	1289	1321
Other Income - Net	43	41
Total Income	1332	1362
EXPENSES EXCLUDING INCOME TAXES	1239	1209
EARNINGS BEFORE INCOME TAXES	93	153
INCOME TAXES-Current and Deferred	8	63
EARNINGS BEFORE SPECIAL ITEMS	85	90
SPECIAL ITEMS		
Tax Benefits from Reading Company Stock Sale	3	5
Net Gain Resulting from B&O Tender Offer	3	-
EARNINGS FOR THE PERIOD	$91	$95
EARNINGS PER SHARE		
Fully Diluted		
Earnings before special items	$4.39	$4.45
Earnings for the period	$4.71	$4.71

"Because of the national recession, the volume of general freight moved was below the year before. Even with the support of higher freight rates, revenues were seven percent below the 1974 figure. Total tonnage moved was close to 1974's level. Aided by increased movements to Tidewater and rate increases, revenues were up over five percent."

* * * * *

"Other income was about even with 1974. The largest item in this category is interest from short-term investments. Also included are income from property rentals, royalty payments on coal produced by mining companies operating on Chessie coal lands, gains on sales of real estate and gains on the re-acquisition of bonds to sinking fund purposes."

"Continuing control over costs was exerted. Nevertheless, railway operating expenses were higher than the year before. This resulted mainly from our program of increased expenditures for roadway improvement. It also reflected higher labor costs and inflationary rises in the price of materials and locomotive fuel."

* * * * *

"Although earnings were down about four percent, dividend payments to shareowners were amply covered. Total dividends paid were $39 million, compared to $33.5 million in 1974."

* * * * *

"The long-range outlook for coal is for continued growth. However, production and use in Chessie territory were affected by a number of limitations in 1975."

"Large stockpiles of coal that had been built up by industry before the 1974 coal miners' strike lingered on into 1975. The weather was generally milder than normal, reducing fuel use. So did the lower levels of industry activity caused by the business recession. Steel mill output was off. Energy conservation practices by industry and the public cut electric power consumption. Lastly, a wildcat walkout by coal miners, mainly in West Virginia, took place in the third quarter."

* * * * *

"A sure harbinger of growth in coal traffic on Chessie System is the continuing development of new coal mines along our lines. Some of the largest of these are on Chessie's own coal properties. Chessie owns or controls through subsidiaries over 700,000 acres of coal land. Of this, approximately two-thirds is leased to mine operating companies."

"Seventeen of the new mines being planned or built just starting up on Chessie System averaged at an annual production capacity of at least one million tons each, for a total of 25 million tons. One is rated at 5 million tons, and another at 2.5 million tons and two more are rated at 2 million tons each. All seventeen are scheduled to begin production in the next three years."

"Ten will provide metallurgical coal exclusively. Two will furnish coal for electric power generation. The remaining five will be able to supply both 'met' and 'steam' coal."

"Immense as is this increase in volume, a vast amount of coal remains for development. The assured and probable reserves in the areas of West Virginia, Kentucky, Maryland, Pennsylvania and Ohio served by the Chessie coal trains amount to nearly 19 billion tons. More than half of this coal has a sulphur content of one percent or less, adding greatly to its value. Even at the present high rates of production, it is enough coal to last for two centuries."

* * * * *

"Even in a year of recession, a number of major new industries began operations on Chessie lines. Nearly $13 million in future annual merchandise revenues are expected from the 111 industries that started up operations. While fewer in total number than the plants opened in 1974, the revenues they contribute to Chessie freight business should be slightly greater."

"Included among the new facilities are a paper warehouse of Pittsburgh Press, at Pittsburgh; an auto parts manufacturing plant of American Motors Corp., at South Charleston, West Virginia; a chemical processing plant of Beker Industries of Canada, at Courtright, Ontario; a corrugated paper mill of Virginia Fibre Corp., at

Riverville, Virginia; a heavy crane assembly plant of FMC Corp., at Lexington, Kentucky; and a warehouse of S. S. Kresge, at Plymouth Michigan."

* * * * *

"Most of the $260 million Chessie spent for capital expenditures in 1975 went for 6,700 new hopper cars. They were the first received of the 16,000 coal cars Chessie ordered earlier to meet future increases in coal traffic. Of the remaining 9,300 cars, 6,700 are scheduled to be received this year."

"Also delivered to Chessie in 1975 were seventy-two diesel locomotives, of 3,000 horsepower each. One hundred sixty-five merchandise cars of various types and ninety-seven all-steel cabooses also started their runs on Chessie rails."

* * * * *

"While the rail-laying forces were setting records, other Chessie track teams were busy. Theirs was the less glamorous, but no less important, task of regular roadway maintenance. In the year they replaced 1.2 million crossties (more than enough to build a new railroad from Cleveland to Baltimore). In addition they replaced 1.3 million tons of ballast, the crushed stone on which the ties and rails rest."

* * * * *

"America's network of railroads was woefully overbuilt during the last century."

* * * * *

"For example, 116 miles of the Western Maryland were built at the start of this century through the valleys of the Potomac and Youghiogheny Rivers in Maryland and Pennsylvania. The line closely paralleled the competing main line of the B&O, just across the river."

"By abandoning the Western Maryland line, Chessie gained more efficient, economical operation. Both WM and B&O use the B&O tracks."

"Chessie has offered to donate to a conservation group a sixty-mile stretch of the abandoned right of way for use as a bicycling and hiking trail. It traverses one of the wildest and most spectacular areas in the Alleghenies. It would open up to the public rugged scenic vistas once viewed only by train crews and outdoorsmen daring enough to ride the wild river."

* * * * *

"As part of the new Richmond facility, Chessie also inaugurated the Virginia Dispatching Center. It is the first of its kind on the system."

* * * * *

"Increased tonnages are expected to move in the months ahead as the pace of the nation's business improves. New and expanded electric power plants and new coke-making facilities at major steel mills will add their demand to that from present facilities."

* * * * *

"Since 1971, Chessie System has obtained permission to close down 619 miles of little-used branch lines with 760 miles of track. It now awaits approval of the Interstate Commerce Commission to abandon another 1,066 miles. Chessie has also removed sections of multiple mainline trackage no longer needed, adding up to a total of 212 miles."

* * * * *

"Construction of a $7.8 million microwave network across the Chessie rail system will be completed in the first half of this year. When it goes into full operation, it will add a vast new dimension to the railroads' communications."

* * * * *

"Chessie System's tentative plans to extend its lines eastward into additional market areas were described in a special report to shareowners dated November 28."

"As stated in that report, if certain conditions were met, Chessie would purchase selected lines, equipment and facilities of three eastern railroads in reorganization. The acquisition would be part of the restructuring of northeast railroads under the Regional Railroad Reorganization Act of 1973."

* * * * *

"Chessie Resources, Inc., the real estate, forest resources and investment subsidiary of Chessie System, Inc., is moving ahead on several fronts to broaden Chessie's income sources to compliment rail earnings. Major real estate developments are planned for several non-rail properties in metropolitan areas such as Chicago, Norfolk, Richmond and Washington, D.C."

"The individual projects vary, with a wide range of commercial, residential and office uses contemplated in joint-venture arrangements with developers and investors. Announcements on at least two of these projects are expected early this year."

"Chessie Resources is a partner in three successful oil and gas drilling ventures along the Gulf Coast of Texas and Louisiana. Reserves discovered have already substantially exceeded Chessie's investment and will provide sustained income in future years."

* * * * *

An obscure news item located on the twenty-first page of the Herald Dispatch in Huntington dated Monday, October 20, 1975, confirmed a drastic change in management:

"Baltimore (AP) - Hays T. Watkins has been elected President of the Chessie System, Inc. following the resignation of John W. Hanifin."

"Watkins, 49, also serves as Chairman and Chief Executive Officer of the corporation and its principal subsidiaries, the Chesapeake and Ohio Railway Company and the Baltimore and Ohio Railroad Company."

"A Chessie spokesman announced the changes yesterday. He said Hanifin, who is 54 and a native of St. Albans, West Virginia, planned to enter the energy field with emphasis on the coal industry and those plans would be disclosed soon."

* * * * *

One Monday morning in the fall, our office received the distressing news that Ross Conlin had died on the previous Sunday of a heart attack while dressing for church. This was a hard blow for me because I really liked him and had tried to keep in touch even though he had been transferred out of the department. We had dinner together a couple of times and it was obvious that he was severely disappointed to have been transferred out of our department although he never made a complaint. I attended the funeral held in Potomac, where he moved when transferred to the government liaison office in Washington. It was a very sad occasion and I grieved in the loss of my friend.

James LaRosa is a coal operator on the B&O from Clarksburg, West Virginia. He converted a surface mine into a mall site on Bridgeport hill. He is owner and builder of the innovative Pete Dye Golf Club out past the Meadowbrook Mall.

My right-hand man, Jim Lilly, got discouraged and asked me to get him a raise which I was unable to accomplish even though I knew he deserved it. He subsequently resigned and was able to secure a more lucrative position with Amax Coal Company, one of the largest western mining companies. They permitted him to continue living in Huntington and search for coal properties or mines they could purchase in the Appalachian field.

☆

JAMES D. LaROSA

JOSEPH L. McQUADE

J. HILARY KELLEY

CARL DEL SIGNORE

1976 - DISAPPOINTMENT IN ACQUISITIONS

The summary of earnings issued in the flash annual report follows:

Summary of earnings Years Ended December 31 (In Millions)	1976
Income	
Merchandise Revenues	$780
Coal, Coke and Iron Ore Revenues	553
Other Freight & Passenger Revenues	82
Total Railway Operating Revenues	1415
Other Income, Net	39
Total Income	1454
Expenses:	
Payrolls	590
Materials, Supplies & Other Expenses	411
Depreciation & Net Retirement Losses	65
Total Railway Operating Expenses	1067
Property & Other Taxes	121
Equipment & Joint Facility Rentals	87
Interest	61
Total Expenses	1336
Earnings Before Income Taxes and Extraordinary Items	118
Income Taxes - Current & Deferred	16
Earnings Before Extraordinary Items	102
Extraordinary Items	-
Earnings for the Year	$102
Earnings Per Share	
Fully Diluted	
Earnings before extraordinary items	$5.19
Earnings for the Year	$5.19

Hays Watkins had changed his title to Chairman and President. His letter to stockholders in the flash annual report was mostly upbeat as follows:

"For the first time in its history, Chessie earned over $100 million - to be exact, $102 million, an increase of 12 percent over 1975 earnings and up seven percent from the previous record of $95 million set in 1974."

"Earnings per share of $5.38 compared with the $4.91 reported last year and the previous record of $5.33 in 1974."

"Strong at the year's end were cash and short-term investments of $195 million and working capital of $45 million."

"In anticipation of those fine results and reflecting their confidence in the future, in October the Board of Directors increased the annual dividend rate by more than ten percent to $2.32, again a new record."

* * * * *

"As the year began, Chessie was deeply involved in what was, for us, to be a frustrating, disappointing experience - the reorganization of the Northeast railroads. As you know, the reorganization was brought about by the bankruptcy of the mammoth Penn Central system and smaller roads in the region."

"In 1973, Congress passed the Regional Rail Reorganization Act, which eventually led to the creation of the Consolidated Rail Corporation, or ConRail, as a government-financed successor to the insolvent lines. Chessie was offered the chance to acquire 2,200 miles of lines of the Erie Lackawanna, Reading and Penn Central railroads. We were prepared to pay $54 million in cash for the properties and invest an additional half-billion dollars in putting them back into shape."

"The acquisition would have extended Chessie's lines into important markets in the East. After many months of planning and negotiations, however, we could not reach an agreement on labor conditions by the deadline in the Federal stature. The lines we wanted became part of ConRail. As if that were not frustration enough, the first quarter held more disappointment. One was January's bad weather, which hampered mining operations in the coal fields, as well as our system-wide railroad activities. There was a rash of wildcat coal miner strikes that further disrupted coal production. World-wide recession reduced demand for U. S. coal, cutting deeply into our export shipments."

"Lastly, the U. S. economy failed to revive as quickly as many had expected it would."

* * * * *

"Earnings plunged to $5.8 million, or thirty-one cents a share, nearly two-thirds less than the $16.1 million, or eighty-seven cents a share, earned in 1975's first quarter."

"The economy was picking up as the second quarter opened. As the industries along our lines became busier, shipments of steel and iron ore increased substantially, and loadings of automotive products continued high."

"Loadings of metallurgical coal rose once more to supply the nation's steel mills. Electric utilities called for more steam coal, and that part of the traffic picture brightened, too."

"Also revenues were helped by an increase in freight rates which became effective in March."

"Chessie reported record second-quarter results, with earnings of $37.3 million, or $1.96 a share."

* * * * *

"Automotive carloadings held up well, despite the strike at the Ford Motor Company, but as business generally slowed down, so did our shipments of most other merchandise commodities."

"For four weeks in July and August, coal production in Chessie territory was seriously disrupted by a widespread miners' walkout. Once the strike ended, however, the mines resumed production quickly, in contrast to the delay after the 1975 stoppage. Coal shipments again began to flow in volume over our lines."

"Because of the fast recovery of coal traffic and a continuing effort of cost control, the third quarter was an excellent one. Earnings of $22.2 million, or $1.17 a share, were the second highest for any third quarter."

"The fourth quarter is usually a strong business period for Chessie's railroads, but 1976 turned out different in this respect, too. The economy remained dormant, and carloadings slackened."

John T. Collinson *
President, C&O and B&O

John T. Ford *
Executive Vice-President
Commercial

Hewitt Biaett
Coal Business Development, C&O/B&O
Native of Arkansas; earned law degree from University of Arizona; worked as commerce attorney; named to present position in January 1966; father of two boys and a girl; wishes he had time to golf, bowl, fish, boat, play bridge.

Robert C. McGowan
Vice-President
Administration
President
Chessie Resources, Inc.

Kenneth C. Morris
Vice-President
Engineering and Maintenance

Richard G. Rayburn
Vice-President
Transportation

John W. Snow *
Vice-President
Government Affairs

"We ended the quarter with respectable earnings of $37 million, or $1.94 a share. Those good results, while not up to the previous year's mark, were achieved through the continued hold down of expenses and the helpful effects of another freight rate increase granted in October."

"Looking back over the year, our railroads performed commendably in moving the business that was available. Chessie maintained its position as the nation's largest transporter of coal and a leading carrier of merchandise freight."

"To insure that Chessie railroads will remain leaders, we spent $232 million for capital improvements and put $431 million into maintenance of roadway and equipment."

"Our non-rail subsidiaries increased their contributions to the parent company's earnings, and we foresee continued growth ahead for them."

"A three-man executive office for the direction of the company and its subsidiaries was established early in the year. To share the executive function with me, John T. Ford and John T. Collinson were elected executive vice presidents. Mr. Ford's particular responsibility is coordination of the commercial areas and overseeing the subsidiary railroads; Mr. Collinson's particular responsibility is operations and maintenance. I am pleased to say that this arrangement has worked out very well."

"Two seasoned newcomers stepped into important vice-presidential posts: Roland W. Donnem, Law, and Jerry E. Gobrecht, coal traffic. Mr. Donnem had been vice-president, secretary and general counsel of Standard Brands, Inc. Mr. Gobrecht came from the Louisville and Nashville Railroad, where he was in charge of coal traffic."

"Also standing out as an event of future significance was our acquisition of a 22 percent interest in the National Mine Service Company, of Pittsburgh. I was elected to the board of directors of this well-managed, growing concern at its annual meeting in November."

"Industries continues to build and expand factories along our lines, attracted by good sites and Chessie service. A large number of coal mines are also being built and enlarged in our territory; in five years we see a rise of 40 percent in coal traffic coming from these mines."

Hays presented additional information at the annual meeting:

"The 78 million tons of coal that Chessie carried away from mines in their territory to steel mills, power plants and other industries were slightly below 1975's volume."

"Total coal handled, including shipments received from connecting railroads, amounted to 95 million tons."

"Predictions of a record year of steel production proved optimistic and the substantial growth Chessie had looked for in shipments of metallurgical coal did not materialize. Shipments of steam coal to electric utilities, while high, showed no growth over those of the year before. Movements of coal for overseas export were down from 1975, a mirror of the recession's impact abroad as well as in this country."

"A rise of 40 percent in Chessie's originated coal traffic is projected over the next five years, a result of more than 100 mines being built or expanded in our territory. Twenty-two of these mines are particularly large, with an annual productive capability each of one million tons or more."

An eastbound WM train climbs upgrade over a trestle near Smithburg. The third unit is just coming into view. Roger Cook.

"Eight of these large mines will be operating this year, including one with an annual production capacity of five million tons. Five more will start up next year, and four are expected to come on stream in 1979. The remainder should begin production between 1980 and 1982."

"Helping Chessie's coal originations in 1976 were three major mines that entered initial phases of production. In full operation, they will have a total productive capacity of four and a half million tons."

* * * * *

"In 1976, production of coal in the United States rose only 2.6 percent, from 648 million tons the year before to 655 million. At the same time, the nation increased its consumption of foreign oil to approximately twice the volume at the time of the 1973 embargo."

"The United States' estimated need for a billion tons of coal a year by 1985 may not be met because of obstacles in the path of greater development and use, according to Dr. Thomas V. Falkie, director of U. S. Bureau of Mines."

* * * * *

"It is interesting to note the report in the Wall Street Journal of December 10 that the Soviet Union has chosen coal as its chief source of power for the years ahead. That country aims at an annual production of 800 million tons a year by 1980."

"A rise of over 15 percent in loadings of vehicles and automotive products was the strong point in the year for Chessie's merchandise freight traffic. Shipments of other commodities matched the tempo of the economy."

* * * * *

"In September, Chessie System placed in service an all-piggyback train over B&O lines between Philadelphia and St. Louis, a market for which Chessie's research indicated a good profit potential."

* * * * *

"Of the 200 merchandise freight trains moving daily over Chessie lines, nearly thirty are 'run-through' trains."

* * * * *

"A brand new Chessie run-through will begin operating this week in cooperation with the Seaboard Coast Line and the Clinchfield Railroad. It will run between Savannah, Georgia, on the SCL, and Russell, Kentucky, a major Chessie classification point, where other Chessie trains will pick up its cars for final delivery."

* * * * *

"In 1976, 125 new or expanded facilities, a record number, were located on Chessie-served sites. These are expected to produce nearly 86,000 future carloads annually and $31 million in additional annual revenues."

* * * * *

"The Chessie railroads' ability to serve their customers, present and future, was strengthened in 1976 by a $232-million capital improvement program, second highest in the company's history."

"Chessie's freight car fleet, with 131,000 units, is one of the largest in the industry. Over half of this steel-wheeled armada consists of coal hopper cars, of which an additional 4,920 were acquired in the year at a cost of $118 million."

"These new hoppers constituted the second major segment of the half-billion dollar commitment for 16,000 coal cars of 100-ton capacity which Chessie announced in 1974. The railroads' own shops at Raceland, Kentucky, were assigned the construction of 10,000 of these cars, with the remainder coming from commercial carbuilders."

"Also acquired were 3,108 cars for transporting merchandise freight, the source of 55 percent of Chessie's railway operating revenues. The bill for these came to $82 million."

* * * * *

"Chessie's own carbuilding shops constructed 162 gondolas of 100-ton capacity for moving coils of sheet steel. These heavy coils are easily damaged by bumping, denting or corrosion."

* * * * *

"For the construction industry, Chessie put into service 238 uniquely designed light-weight hopper cars for moving aggregate, or crushed stone. And in another forty-eight enclosed tri-level cars, called 'Rail-Paks,' new automobiles are riding to market with greater protection from vandalism."

"Construction of a $56-million automated classification terminal at Cincinnati will begin early this year."

* * * * *

"The Cincinnati terminal will enable Chessie to eliminate eight of fourteen yards now scattered over a twenty-five radius in the area."

"A nine-mile track project in the Johns Creek area of southeastern Kentucky will provide better service to new and existing coal mines in the region and substantially reduce operating costs for Chessie."

* * * * *

"Covering 1,000 route miles, Chessie System's new microwave communications network will generate savings through reduction of leased communication lines and other telephone company charges. It will also greatly increase the communications channels available to Chessie people."

* * * * *

"In 1977, Chessie System plans to invest another $160 million for capital improvements. Of this, $110 million will go for 3,800 additional freight cars and thirty-five diesel-electric locomotives. For fixed plant improvements, $50 million will be spent."

"In 1976, Chessie System spent $202 million for maintenance of its roadway and structures."

"Over 500 additional miles of continuous welded rail were installed. That pushed Chessie's total mileage of this jointless, smoother-riding, longer-lasting track close to the 6,000-mile mark. That is more than half of all the main line service routes on the system. Chessie continues to lead the industry in welded rail installation."

"In addition, 1,343,000 crossties and 1,405,000 tones of ballast were replaced, both increases over 1975. Re-surfacing of the roadbed was done on 2,900 miles of line."

* * * * *

"Chessie has identified 2,010 miles of track as being redundant and has applied to the ICC for authority to close them down. However, due to the slowness of the statutory procedures involved, only 772 miles, or 38 percent, have been approved by the Commission and the courts, so far."

* * * * *

The present NRC administrative team consists of Executive Vice-President Kirkpatrick; Director J. A. Hunt; President E. V. Bowman and Director Sparks in the front row and Donald M. Reedy, Vice-President in charge of Planning and Development; Secretary-Treasurer O. T. Carr; General Merchandise Manager Norville L. Ellison and J. M. Wade, Consultant in the back row.

"Income from real estate leases and sales, oil and gas ventures and royalties, as well as development of timber lands, increased substantially. More important, projections call for resource activities' contribution to the parent company's earnings to double in five years."

"The White Sulphur Springs Company, which owns and operates the Greenbrier resort hotel, reported patronage at an all-time high."

* * * * *

"Remodeling and expansion of the golf club was completed in the year."

* * * * *

"A new coffee shop and ice cream parlor was opened for those preferring a light lunch or between-meal snacks. The hotel's shop arcade on the lower lobby level was given a completely new appearance. Individual shops were enlarged or rebuilt, new facades were put up and the long hallway was completely carpeted."

* * * * *

"Chessie System, Inc. acquired 22 percent of the outstanding shares of National Mine Service Company, of Pittsburgh, in October."

"The company's products and services cover the whole range of systems, machines and equipment used to mine coal and other minerals in this country and overseas."

* * * * *

I never got to know Allen Pack until he succeeded Paul Morton as President of Cannelton Coal Company. Paul was the last in a long line of Mortons who were prominent in the coal industry of West Virginia. Brother Quinn was the long time Executive Secretary of the Kanawha Coal Operators Association in Charleston. Allen was a third generation coal miner and later retired from the Presidency of Cannelton.

Jimmy Carter defeated Ford only to prove himself not very helpful to the business community.

One of the perks of my job besides membership in the local country club and Rotary clubs was one of the company memberships in the local Chamber of Commerce. I was notified that I had been placed on the Transportation Committee and that a meeting of that body was called for the next week. I showed up to participate in a discussion about the fact that the city had made Tenth Street one way south and Eleventh Street one way north in the downtown area mainly to accommodate a new bank building. I made a vigorous argument that one way streets were a pain in the neck to motorists especially visitors and that the streets should be changed back to two-way traffic. I was glad to learn later that two-way traffic was restored.

During the year, the operating department decided that the new coal business being generated on the Johns Creek Branch in Pike County, Kentucky could no longer handle the coal traffic and therefore wanted to build a new eight-mile outlet to the mainline on Levisa Fork. This was the extension that I had recommended to management in order to replace the line which was removed to make way for the Fishtrap Dam. All of our projections for more and better reserves were now paying very handsome dividends and several new mines had come on stream. Seldom does one have the experience of seeing his projection turn out so well for new business and greater profits.

On Thursday, August 26, a disgruntled employee arrived on the 36th floor of the terminal tower in Cleveland. The newspaper in Huntington displayed the story of the front page the next day:

"A Huntington man who for several years has campaigned against the Chessie System for GI benefits last night released nine hostages he held at gunpoint for nine hours in the railroad's headquarters here and surrendered to police.

No one was injured.

Ashby Leach, 30, of 2935-1/2 3rd Ave., said taking the hostages 'appeared the only way left available to me to bring about moral justice' in his quarrel with Chessie's policies on veteran's benefits.

Leach had said he would release his hostages after he saw his demands on national television. He kept that promise after all three television networks ran news stories on Leach's demands on the evening news.

'Obviously, now the next move is up to him,' an ABC reporter commented in his televised report.

Minutes later the hostages, including former Huntington resident, Vary P. Mengle, Jr., were released. After the newscasts, Leach surrendered to police on the 36th floor at 7 p.m. After a short scuffle on the street, he was placed in a police car."

* * * * *

"The downtown area of this city is practically vacated after 6 p.m. on any other Thursday. Yesterday was different.

An estimated 500 persons-from scruffy black children to businessmen in three-piece suits-converged around the Terminal Tower as Ashby Leach of Huntington, held 13 persons at gunpoint and law officials at bay for nearly nine hours.

Blue-shirted police barked half-hearted commands, asking people to move away from the front door of the building. Police cars were nearly outnumbered by remote trucks from television and radio stations."

* * * * *

"Leach said he wanted his demands broadcast on network television. The hour--6:30--came and went. The local ABC affiliate reportedly carried his demands."

* * * * *

"At 7 p.m. sharp a cordon of police stormed through the lobby toward the outside of the building, pressed in on all sides by spectators and newsmen."

"Leach, handcuffed in the center of the cordon, towered over the policemen. His curly white hair, flecked with black, made him look older than his 30 years. he was dressed in a white T-shirt and trouser."

"Leach was charged with 13 counts of kidnapping and one count of criminal damaging, the latter charge in connection with shots fired into the ceiling and into a radio he ordered brought to him to monitor cooperation of authorities."

* * * * *

"Leach, a Navy veteran who received psychiatric treatment after returning from Vietnam, apparently left Huntington for Cleveland early Wednesday morning, his wife said."

* * * * *

"The drama began about 10 a.m. yesterday when Leach, armed with a sawed-off shotgun and a pearl-handled pistol, walked into a reception area on the 36th floor of Terminal Tower in downtown Cleveland, the FBI said. He entered the offices through a private elevator and pulled a shotgun from under a white raincoat, authorities said.

'He fired at the ceiling and said he wanted to talk to someone about his benefits,' said Milt Dolinger, a Chessie assistant vice president. Later, he fired a second shot into a radio.

Throughout the day, Leach communicated his demands through both police and the news media."

* * * * *

"But Leach's demands were not broadcast until hours after Chessie officers agreed to submit to them.

In Cleveland, Chessie public relations vice president Howard Skidmore read to reporters a letter signed by R. C. McGowan, Chessie vice president for administration."

* * * * *

"Leach had worked for Chessie for nearly a year as an apprentice machinist when he was laid off in August 1971, his wife, Linda, said. He also worked there briefly in 1974."

* * * * *

"From the information given The Herald Dispatch by family members and friends, it appears Leach's life of bad luck and troubles began more than 10 years ago when, at the age of 19, his first wife was killed in a fire at the couple's apartment at 419 Rear, 26th St."

* * * * *

☆

1977 - A FRUSTRATING YEAR

Hays Watkins letter dated February 15, 1978 explained the Summary of Earnings which accompanied the flash annual report.

SUMMARY OF EARNINGS Year Ended December 31 (Millions)	1973	1972
Income		
Merchandise Revenue	$840	$780
Coal, Coke and Iron Ore Revenues	561	552
other Frt & Pass Revenues	91	82
Total Railway Op. Revenues	1493	1415
Other Income, Net	46	39
Total Income	1539	1454
Expenses:		
Payrolls	645	590
Materials, Supplies & Other Expenses	446	411
Depreciation & Net Retirement Losses	70	65
Total Railway Op. Expenses	1161	1067
Property & Other Taxes	131	121
Equipment & Joint Facility Rentals	83	87
Interest	67	61
Total Expenses	1442	1336
Earnings Before Income Taxes and Extraordinary Items	97	118
Income Taxes - Current & Deferred	19	16
Earnings Before Extraordinary Items	78	102
Extraordinary Items	10	-
Earnings for the Year	87	102
Earnings per share		
Fully Diluted		
Earnings before extraordinary items	$3.77	$5.19
Earnings for the year	$4.27	$5.19

The Chairman's letter is excerpted as follows:

"As you know, 1977 was marked by a first-quarter loss of $7.4 million, caused primarily by the worst winter on record. In the summer, we were racked by ten weeks of coal strikes centered in Chessie System territory. The year closed with the mine workers going out on an official strike on December 6."

"The strike continues in its eleventh week as this letter is written. On February 12, the Bargaining Council of the United Mine Workers Union overwhelmingly rejected a contract agreed upon by negotiators for the union and the coal mine operators after four months of talks. How much longer the strike will last is not now known"

* * * * *

"The strike's devastating impact on Chessie's coal traffic is certain to generate a first-quarter loss in 1978, one considerably worse than in last year's first-period deficit."

"The more than 400 mines presently served by Chessie railroads are being joined by new mines each year and many more are being developed or planned. As a result, the expansion of production capacity - the maximum potential output of all mines in Chessie territory between now and 1982 should be right in line with the increase in national coal production called for by President Carter. This excellent outlook, moreover, does not take into effect the future mine development certain to materialize in Chessie's coal region."

* * * * *

"Last fall, following the wildcat strikes, Chessie loaded coal at an annual rate of 100 million tons."

"Chessie's emphasis on its coal operations takes nothing away from the importance of its merchandise traffic. Our railroads' location in the industrial concentration in the Northeast and their role in supplying service to basic industries positions them as a major transporter of merchandise freight. Merchandise carloadings, in fact, are the source of more than half of Chessie's railroad revenues, and they too, offer potential expansion."

* * * * *

"Railroads and the unions which represent our employees are presently negotiating new three-year contracts. The bargaining is more sensitive than usual because the industry wants the unions' cooperation in changing certain work rules and pay formulas to offset some of the rising wage costs with increased productivity."

* * * * *

"Dr. Kimpton, a director of Chessie System, Inc., C&O and B&O, died on October 31. He was renowned as an educator, scientist and business executive. During World War II, he was chief administrative officer of the atomic development project at the University of Chicago, which he later headed, as chancellor, from 1951 to 1960. He then became a senior executive of Standard Oil Company (Indiana)."

"Mr. Cowan, who died on June 13, was director of B&O and Western Maryland. A veteran, well-loved railroader, he started with C&O in 1936 as a brakeman and rose to senior vice president of C&O and B&O, the post he held at retirement in 1972."

"Chessie's management team was augmented in the year by the election of two new vice-presidents. Richard G. Rayburn was named vice-president for transportation and John W. Snow was appointed vice-president for government affairs, a new post."

"Mr. Rayburn, 51, joined the B&O in 1950 as a technical trainee after graduating from Purdue University with a degree in civil engineering. He had been assistant vice-president for transportation since 1971."

"Mr. Snow, 38, is an attorney and educator and before joining Chessie held a number of high posts in the Department of Transportation. He is a graduate of Kenyon College and holds a doctorate in economics from the University of Virginia."

* * * * *

"We are doing everything we can to minimize the influence of a first-quarter loss of the year's outcome. No opportunity to capitalize on an advantage will be overlooked. Every prudent means of reducing or controlling expenses is being undertaken and will be pursued."

The President's letter was augmented with captioned pictures to give stockholders additional information:

"In 1977, Chessie System, largest transporter of coal in the nation, saw its coal traffic affected by uncontrollable forces - in the first quarter, the worst winter weather on record, widespread wildcat strikes for ten weeks in the summer, and, in the last three weeks of the year, an official walkout by members of the United Mine Workers Union which continued well into 1978."

* * * * *

"For the year, coal movements totaled 91 million tons, of which 78 million came from the mines served directly by Chessie. That compared with the 95 million tons overall moved in 1976, of which 78 million tons originated in Chessie territory."

"At Baltimore, where winter is usually mild, Chessie workers had to use air hammers to break up large chunks of coal clogging car-dumper hoppers."

"Chessie's answer to future freeze-ups include Galloping Gertie, designed and built by the railroads' own engineers and shops. The machines' vibrating prongs, lowered into coal cars, reduce frozen coal to manageable size. They were part of a $3.5 million anti-winter equipment program at Baltimore and Newport News, Virginia."

* * * * *

"Among major mines that opened on Chessie System in 1977 was Amco Steel Corporation's Number 10 Mine at Sundial, West Virginia, on the C&O. It will supply Armco with one million tons annually of metallurgical coal."

* * * * *

"Capital expenditures of $144 million in 1977 brought Chessie's investment in physical improvements over the last five years to $727 million. Acquired at a cost of $91 million were thirty-five diesel-electric locomotives from General Motors and 2,484 freight cars, including 1,352 100-ton hoppers, 355 heavy-duty boxcars, 313 coil steel cars, 369 special multi-level automobile carriers and 95 other merchandise freight cars. An additional $47 million went for property improvements and additions."

"Roadway maintenance entailed installation of 515 miles of welded rail, replacement of 1.4 million crossties and 1.5 million tons of ballast, as well as re-surfacing of over 3,000 miles of line."

* * * * *

"Celebrating the 150th anniversary of the Baltimore and Ohio, the nation's first railroad, Chessie System operated a special steam train on forty-six tours through eight states between April and October, 1977."

* * * * *

"Ground was broken in November for Freemason Harbor, a vast development project in a historically-significant area of downtown Norfolk, Virginia."

* * * * *

"Chessie and the Carr Company have formed another joint venture, called Camden Station Associates, for development of surplus B&O property behind the historic passenger terminal in downtown Baltimore."

* * * * *

"Chessie, a leading supplier of transportation to the automotive industry, moved 346,400 carloads of motor vehicles and parts in 1977."

* * * * *

"A Chessie train helped dedicate the new four-mile rail line and yard built for the Volkswagen of America plant in southwestern Pennsylvania, expected to begin production in April, 1978."

* * * * *

"Chessie's share of the piggyback market was enlarged in 1977 with introduction of an all-trailer train between the East Coast and Chicago. A similar train, operating between the East Coast and St. Louis, inaugurated service in 1976. Both offer second-morning delivery and set excellent performance records in the year."

* * * * *

Our office had been sponsoring a golf outing at the Greenbrier in June every year since the late fifties to entertain the operators in the New River and Greenbrier field where many of our mines were jointly served by the Virginian or New York Central Railroads. Joe McQuaid had been an operator on the B&O and WM lines for a long time, but he came back to southern WV and took over an old mine on Paint Creek. He started coming to our party and I got to know him quite well. Besides producing lots of coal, he was noted for raising fifteen children with his wife, Marion. This couple has become well known for a variety of causes and are still active in Oak Hill, West Virginia.

Besides Del Signore, there were several coal operators in the Western Maryland coal fields that were very good traffic customers. James T. Watson operated the oldest coal mine at one site in West Virginia along the North branch of the Potomac River. Lawrence Streets, President of Allegheny Mining, and Dick Swinabart were consistent producers who contributed coal traffic for the railroad. Marlin Sherbine was manager of the P.B.S. Coal mines near Somserset, Pennsylvania.

I came off the road from a trip to my office in Thomas, West Virginia to learn that the statue of Colonel Huntington was being removed from in front of the Huntington passenger station and moved to a new tourist development being built at the site of the old B&O depot near the river. I was appalled that anyone would even consider removing the only viable landmark in the city just to enhance a dinky little development to be called Heritage Village. I still hope that a future Mayor will have the courage to restore the magnificent statue to its rightful place.

I was reading the proxy information that came out in February with the announcement of the annual meeting and I noticed that under the heading of executive compensation, a statement revealed that 70 officers were on a bonus list. These officers all received a bonus in addition to their salary depending on the earnings of the Company at the end of the year. I was flabbergasted to think that there could be 70 officers in the company making a bigger contribution to earnings than my own office. After thinking about the matter for a couple of days, I decided to request promotion, salary increase and inclusion on the bonus list. On my next visit to Cleveland, I was advised by our Vice President Henshaw that he could not help me because I was making as much as he did as an AVP. I requested that he go with me to petition higher authority and on my next visit to Cleveland, I was advised to go see Norman Halpern. I had known "Norm" from way back when I was on his communication committee, but he had gone up in the world to Senior Vice President. He asked me several questions which he already knew the answers and ended the interview by expressing his opinion that promotion was out of the question. Several days later, I was notified by the personnel department that my name had been added to the bonus list. This seemed to me to prove the old adage that if you don't look out for yourself, no one else will.

Bruce Bevins was forced off the highway in his company car by a large coal truck but fortunately suffered no injuries. Jeff Conley, a graduate in geology was hired to make mine inspections and is proving to be a good worker. Bruce's accident helped us ward off a management program to convert over to compact cars as an economy move. We argued that the full size Chevrolets and Fords were safer and improved the image of our representatives when dealing with customers out in the field.

My first company car was a black Ford sedan purchased from the Huntington dealer in late 1947 when new cars were still very scarce after the war. It was the same 4-door sedan used by several county Sheriff departments and I got a kick out of seeing other motorists slow down when I traveled. In my 34 year career, I think I had 9 different company cars, all Chevrolets after the first Ford until 1963 when I rated an Oldsmobile as a departmental head. I traveled about 15,000 miles per year in my company cars which adds up to a total of a half million miles. After I retired, I calculated that I had spent over 10 years of my life in motels and hotels.

☆

1978 - ANOTHER PRESIDENT AND MERGER

Although Hays T. Watkins retained the title of President of Chessie System, Inc., John T. Collinson became President of C&O/B&O. A special stockholders meeting was held on February 13, 1979, to approve merger with the Seaboard Coast Line Industries. Hay's letter to stockholders dated March 1, 1979 is excerpted as follows:

"The proposed merger with Seaboard Coast Line Industries, Inc., first announced September 6, 1978, occupied much of our time in the last quarter of the year. At that time, we announced merely the start of studies on the possibility of a merger of our parent companies. Looking back, we have come a very considerable distance in a surprisingly short time. Those studies by joint teams from both companies were completed so swiftly - and thoroughly and accurately, I might add - that it was possible to draw up a plan of merger and present it to the boards of directors just ten weeks later on November 16. The boards unanimously and enthusiastically approved the plan."

* * * * *

"The Seaboard/Chessie merger cause was strongly advanced by an act of statesmanship on the part of Benjamin F. Biaggini, chairman and President of the Southern Pacific Company, the largest holder of Seaboard stock, which had itself sought affiliation with Seaboard."

* * * * *

"Cyrus S. Eaton, 95, chairman emeritus and a valued director of C&O and Chessie for thirty-five years, did not stand for re-election at the 1978 meeting. Joining the board were Nicholas T. Camicia, 62, chairman, president and chief executive officer of the Pittston Company, and Dr. Steven Muller, 51, president of The John Hopkins University and The Johns Hopkins Hospital. John T. Collinson, 52, a fourth-generation railroader who joined Chessie in 1948, was elected president of the Chesapeake and Ohio Railway and the Baltimore and Ohio Railroad. Roland W. Donnem, 49, and Robert L. Hintz, 48, were elected senior vice-presidents. Jerome D. Krassenstein, 47, and Kenneth C. Morris, 54, were appointed vice-presidents."

The annual report to stockholders was in addition to the letter:

"A devastating coal strike, the longest on record, combined with another bitterly harsh winter to cause a first-quarter deficit of $67 million."

"Then in the next three quarters, Chessie made an amazing recovery, strong enough not only to wipe out the deficit, but to generate record earnings in the next three quarters of $137.6 million. Chessie earned more in 1978's last nine months than it ever earned in a full year."

* * * * *

"As it was, 1978 earnings of $70.5 million were exceeded by only four other years."

"Railway operating revenues of more than $1.6 billion set a record, increasing over 1977 by eight percent. The rise was mainly the result of freight rate increases by the Interstate Commerce Commission to offset inflationary cost increases."

* * * * *

"Merchandise freight revenues of $915 million were up $75 million, or nine percent, reflecting the higher rates, a slight increase in carloadings and an improved traffic mix, i.e., more shipments of high-rated commodities, longer hauls and heavier loadings, increasing the average carload revenue."

"Revenues of $583 million from transporting coal, coke and iron ore, historically Chessie's largest single traffic source, represented 36 percent of railway operating revenues and a four percent increase over the previous year. That modest rise, however, must be weighed against Chessie's loss of $10 million in coal revenues during the strike-bound first quarter. Carloadings for that period fell 294,000 short of expected levels."

"Miscellaneous freight and other revenues of $110 million, covering charges for switching cars, shippers' detention of cars, use of piers, etc., were 20 percent above 1977's mark."

"Other Income-Net of $44 million was down slightly from 1977. It includes a number of non-rail income sources: royalty payments on coal and timber produced by companies operating on Chessie-owned land, gains on bonds reacquired, income from property rentals, gains on sales of real estate and interest on short-term investments."

* * * * *

"While Chessie took in more money in 1978 from providing rail transportation and other services, it had to pay out more money in the process. Expenses of over $1.5 billion - an average of $4 million a day - increased $134 million, or ten percent, over 1977."

"The major factors forcing up the cost of doing business were higher mechanical and roadway maintenance expense ($490 million), greater payroll costs and inflationary rises in the prices of materials."

* * * * *

"However, payroll dollars - the combination of direct payroll and fringe benefits - rose to $895 million, up $67 million, or eight percent, over 1977."

* * * * *

"Despite the first-quarter loss and capital expenditures of $232.2 million, third highest in company history, working capital at $33.8 million was up from 1977's year-end mark of $28.7 million. Cash and short-term investments at $102.9 million, while off appreciably from 1977, were still more than adequate."

* * * * *

"The dividend of fifty-eight cents a commonshare, $2.32 annually, was maintained, and shareowners received a total of $45 million in dividends."

* * * * *

"In any twenty-four hour period in 1978, about 800 road crews (more than 4,000 men and women in the full year) were moving that many trains over the main line and branches."

* * * * *

"At the massive complex at Raceland, Kentucky, near Ashland, they built over 2,500 brand-new, 100-ton coal hopper cars."

* * * * *

"In 1978, among the other assignments, those 7,200 railroaders installed nearly 500 or more miles of continuous welded rail (for a new system total of 6,886 miles),

1.8 million new ties and 1.4 million tons of ballast, resurfaced over 3,600 miles of track, repaired 356 bridges and installed 7,496 radios."

* * * * *

"Some thirty track projects like this were underway in the year with a total value of nearly $36 million."

* * * * *

"Chessie has already eliminated 1,004 miles of redundant lines and has applied to the Commission to abandon 434 more miles. Another 664 miles are possible candidates for abandonment."

* * * * *

"Also being appealed is the I.C.C.'s approval of an agreement between Chessie, shippers and state agencies to permit abandonment of the Lake Michigan trainferry service to Milwaukee and, after January 1, 1980, to Manitowoc, Wisconsin, if traffic levels to that port do not rise substantially."

* * * * *

"In 1978, Chessie completed installation of additional computer equipment at Baltimore at a one-time cost of $900,000. It will provide capacity for more data processing applications and will expand current applications, resulting in significant savings."

* * * * *

"The nation's premier carrier of coal and a leader in moving merchandise freight, Chessie railroads transported 3.4 million carloads in 1978. That was down one percent from 1977, mainly reflecting coal traffic lost during the first-quarter strike. Merchandise shipments were about even with the year before."

* * * * *

"Merchandise freight movements exceeded 1.9 million carloads in 1978. The commodity shipped in greatest volume was, once again, automotive-related products, which Chessie trains handled at a rate nearly equal to 1,000 carloads for every day of the year. Chemicals, consumer goods, minerals, metals and food and grain followed in that order."

* * * * *

"In 1978, it acquired 2,522 new coal hopper cars, 785 sixty-foot parts box cars, 320 multi-level rack cars for carrying finished automobiles, 565 covered hopper cars, and 1,596 forty-foot piggyback trailers."

* * * * *

"Plans were announced for a luxury residential community adjacent to The Greenbrier, the resort hotel owned by Chessie System, Inc., at White Sulphur Springs, West Virginia."

* * * * *

"A Chessie joint venture with Western Development Company, a Washington-based firm, will develop a prime 2.5 acre parcel located along the Potomac River in the Georgetown section of Washington, D.C."

* * * * *

"In a joint venture with Chessie, The Glimcher Company, of Columbus, Ohio, plan to commence construction of a shopping center in 1979 on a site near Beckley owned by another Chessie subsidiary."

* * * * *

The Baltimore and Ohio Railroad Museum in the former Mount Clare passenger station and roundhouse in Baltimore, Maryland.

Very early passenger train on display in the B&O museum.

"A computerized inventory system, completed in 1978, is being utilized to develop harvesting plans and growth projections for Chessie's 350,000 acres of timber holdings."

* * * * *

"Chessie moved to diversify within the forest products industry in 1978. Eastern Hardwoods Inc., a new company owned by Chessie Resources, Inc. and J. C. Hamer, broke ground for a sawmill near Beckley, West Virginia."

* * * * *

"The joint venture programs in which Chessie participates drilled 117 wells in 1978, of which sixty-three were completed as capable of producing in commercial quantities. These include twenty-one new field discoveries and forty-one development wells which expand previously discovered fields."

* * * * *

"THE GREENBRIER WILL HOST RYDER CUP IN SEPTEMBER.

The biggest attraction at the hotel in 1979 will be the 23rd International Ryder Cup Matches, from September 11 through September 16. Expected to compete will be the best golfers of the United States and Europe, such as Tom Watson, John Mahaffey, Hubert Green, Jack Nicklaus, Severiano Ballesteros and Nick Faldo. The Matches will be played on The Greenbrier Course, newly redesigned by Mr. Nicklaus for the purpose."

* * * * *

The B&O had a fine museum established near Camden Station in Baltimore. Willis Cook from the public relations department of C&O was in charge of the facility for several years during which time many improvements were made. It is now a world class facility and a must see attraction for anyone interested in the listing of railroads in the USA.

In April of the year I was called to Cleveland for regular consultation. It was a meeting that I really looked forward to because the VP was supposed to retire at the end of June and I felt there was a good chance that I would be promoted and even entertained the idea that I might replace him as Vice President in charge of the coal department. The reality turned out to be a huge disappointment. After conferring in the office on current matters, I was advised that Jack Ford wanted to see me in his office on the 36th floor.

Jack asked me if I wanted some coffee or tea and I declined. Without further ado, he proceeded to outline the new plans which was to send the retiring VP to Huntington to set up a separate office to manage the subsidiary land companies and I was to report to him there.

In my reflections I thought of a saying my mother used on occasion to the effect that "out of sight, out of mind." The fact that I had operated in the field outside of the inner circle really meant that I was a severe disadvantage to be considered for promotion.

Several months later at the Rodeo at the Greenbrier, Jack Ford must have had some sympathy for my situation when he volunteered the statement to my wife that one man alone couldn't do anything about promotion, which seemed to indicate that he had backed me for promotion. At least it made me feel a little bit better to know that someone was on my side.

In June, I got the word that the new Vice President for our department was going to be Jerry Gobrecht, a former Vice President on the L&N Railroad. He had been hired by Henshaw to succeed Burroughs as Coal Traffic Manager when Burroughs requested a return to his old job in Newport News.

☆

1979 - A SPECIAL YEAR FOR EARNINGS

Hays Watkins letter of February 22, 1980 to the stockholders in the annual report was very enthusiastic:

"For Chessie shareowners and for Chessie people, 1979 was a very special year. The company achieved its highest earnings ever - $120.5 million, equal to $6.05 a share. These earnings were 71 percent greater than last year's earnings, when, as you know, we were affected by the longest coal miners' strike in the nation's history."

"Shareowners received a total of $46 million in four quarterly dividends of fifty-eight cents a share. Chessie people were paid more than $1 billion in direct wages and salaries and fringe benefits. Local, state and federal government also shared in Chessie's good fortune, with $45 million in income taxes."

"Revenues from our railroads - the Chesapeake and Ohio, Baltimore and Ohio, Western Maryland and smaller subsidiary lines - also set a record. At $1.86 billion, they were 16 percent above 1978 revenues of $1.6 billion."

* * * * *

"Cash, at $117 million, was equal to the 1978 level after providing funds to extinguish substantial amounts of debt. In addition to a $40 million direct reduction in borrowings, we sold $85 million of redeemable preferred stock and used the proceeds to reduce bank borrowings even further. This financing, which provided a hedge against rising short-term interest rates, saved $1.3 million in 1979 financing costs."

* * * * *

"The B&O, which resumed cash dividends in the fourth quarter of 1978, continued its new policy with two semi-annual payments of $4 per share in 1979. The Western Maryland increased its annual dividend rate on common stock from 80 cents to $1 per share."

* * * * *

"With the formation in 1973 of Chessie System, Inc., we entered a new era of diversification into non-railroad business, while maintaining our position as a leading transportation system."

"Finally, in the last two years, Chessie took its most positive step forward by agreeing in merge with Seaboard Coast Line Industries, Inc., and create a new diversified parent company that will include a coordinated rail system covering most of the eastern and southern parts of the nation. Hearings before the Interstate Commerce Commission, begun in June, 1979, were completed early in 1980 and a decision is expected this year."

* * * * *

"Chessie's board of directors was strengthened in the year by the election of Dr. Mary T. Kimton, noted educator, economist and business consultant."

"Likewise, our executive team was broadened with the election of John S. Lanahan as a senior vice-president of Chessie System, Inc. Mr. Lanahan retains his position as president of The Greenbrier and responsibility for its overall policy. Elected a vice-president of the Chessie railroads was James T. Lyon, with responsibility for taxes."

"On May 9, Cyrus S. Eaton, chairman emeritus and closely involved in Chessie affairs for thirty-five years, passed away."

* * * * *

Further information was included in the annual report.

"Chessie System railroads have been moving longer than any other, since May 24, 1830, to be exact, when the Baltimore and Ohio commenced scheduled service with horse-powered trains over thirteen miles of line out of Baltimore."

* * * * *

"1979 railway operating revenues of $1.86 billion, an all-time high, generated most of the parent company's record earnings. Especially impressive was the dramatic increase in coal, coke and iron ore revenues. At $722.2 million, they were 24 percent over 1978's revenues, which were affected by the longest coal miners' strike in history."

* * * * *

"Merchandise freight revenues passed the one-billion mark for the first time, rising ten percent over 1978's $915.5 million. That increase, due in part to freight rate increases, also reflected strength in movements of grain and chemical traffic. Overall, however, merchandise traffic volume was off by three percent, mainly because of the downturn in the economy in the second half of 1979 and the problems in the automotive, steel and construction industries."

"Expenses rose by 12 percent to $1.8 billion, but the increase could have been substantially greater if Chessie people had not pressed their expense control program and, in the second half, had not smoothly adjusted the levels of railroad operations to the lower volume of merchandise traffic."

* * * * *

"Price escalations added another $46 million to Chessie's fuel bill in the year, and while railroads were authorized freight surcharges totaling 4.7 percent to offset fuel price increases, they did not cover all of Chessie's higher fuel costs."

"Combined direct payroll and fringe benefits also went past a billion dollars, increasing 12 percent over 1978 manpower costs and representing 54 cents of every dollar the railroads took in. The annual average wage for Chessie people increased to $22,500."

"Chessie railroads' physical strength was maintained and enhanced in 1979 through capital expenditures of $141.2 million and maintenance outlays of over half a billion dollars."

* * * * *

"To meet record 1979 grain harvest, Chessie augmented covered hopper fleet with surplus coal cars fitted with hard plastic covers. At shipping season's peak, Chessie moved an average 1,000 carloads of grain per day, a record volume."

* * * * *

"Since 1974, when the trackage reduction program began, more than 1,089 miles of branch lines and sidetracks have been retired, with the permission of the Interstate Commerce Commission, including 85 miles abandoned in 1979. Additional retirements are planned."

* * * * *

"Since 1975, Chessie has been trying to abandon its money-losing Lake Michigan trainferry service which carries cars, passengers and their vehicles between Ludington, Michigan, and Milwaukee, Manitowoc and Kewaunee in Wisconsin. For many years a busy, time-saving short-cut for long-distance rail traffic and motorists, it now generates an annual deficit of $5 million. Improved transit time through Chicago has made that the preferred rail route."

"On December 21, 1979, after numerous appeals from earlier rulings in Chessie's favor, the I.C.C. issued a final decision authorizing abandonment between Ludington and Milwaukee."

* * * * *

"Coal traffic is the largest single source of revenues for Chessie System railroads - 35 percent in 1979 - so it is not surprising that one commodity commands the attention and activity of an entire department."

* * * * *

"In the year, Chessie maintained its leadership as the nation's largest coal transporter, moving 79 million tons from mines served by its railroads. That was a 13 percent increase over 1978 originations. Adding the tonnage received from other railroads for final delivery, Chessie's total coal movements came to 96 million tons, a rise of 13 percent over 1978."

"Chessie's coal traffic will continue to grow, as signalled by the ongoing development of new mines and the expansion of production by existing mines. In 1979, thirteen additional mines were opened, eight mines that were closed more than a year were re-opened and five major track projects were completed to provide more service for expanding mines, The ultimate annual production capacity generated by these developments exceeds 17.5 million tons."

"Chessie maintains the largest coal carrying fleet, numbering 71,600 cars at the end of 1979. The fleet was further strengthened in the year through acquisition of another 888 hopper cars of 100-ton capacity and program repairs to 3,840 existing cars."

* * * * *

"Coal movements of 20.3 million tons through the Chessie export terminals at Baltimore and Newport News were 60 percent over the 1978 volume and were the highest in the seventeen years since C&O and B&O affiliated. The 9.1 million tons moved over the Baltimore pier was a port record. So was the loading there of 84,614 tons of coal into a single vessel on November 23. It is noteworthy that the records were set in spite of damage to the pier's coal-loading machinery caused by Hurricane David in September."

"To assure Chessie's continued ability to maintain a high volume of exports, considerable improvements have been made to the Baltimore pier, aimed at improving winter operations. As additional protection, new freight rates give producers on the B&O and Western Maryland, who would normally ship their coal through Baltimore, a second port outlet through Newport News. Additional expansion and improvements at both the Baltimore and Newport News terminals are being contemplated."

* * * * *

"Total shipments to steel mills, including tonnage received from other railroads, amounted to 31.5 million tons."

"Shipments of Chessie-territory coal to electric utilities fell some 9 percent in the year, down to 34 million tons from 38.1 million tons the year before, largely due to heavy stockpiling after the 1978 miners' strike."

* * * * *

"Two new power plants, one served by C&O, the other by B&O, are scheduled to begin operations early in 1980, and Chessie railroads are already moving coal to them."

* * * * *

"The opening in July of the improved Johns Creek branch line in southeastern Kentucky eliminated severe obstacles, such as steep grades, to providing efficient service to mines in this important coal-producing territory. It involved construction of more than eight miles of new line, five large bridges and a 1,544-foot tunnel at a cost of $13 million. The entire project took only twenty-one months."

* * * * *

"Biggest project in force in the year was construction of Queensgate Yard at Cincinnati, the busiest interchange point on Chessie System."

* * * * *

"Over 550,000 acres of land in West Virginia, Kentucky and Maryland are owned by four Chessie subsidiaries -- Western Pocahontas Corporation, Eastern Pocohontas Corporation, Western Maryland Company and Mid-Allegheny Corporation - and two companies partly owned by Chessie."

* * * * *

"In 1979, new leases and agreements with independent companies were negotiated that will provide for annual minimum royalties of $700,000. Royalties under the new leases are based on a percentage of the sales price of the coal and will assure larger royalty income after operations begin."

* * * * *

"For many years, the primary reason for leasing Chessie coal properties to independent producers for development was to generate additional coal tonnages for the railroads. Now, however, the royalties have become a more productive source of income and are receiving greater emphasis."

"The New River Company is an independently operated coal-mining subsidiary of Western Pocohontas Corporation. It is building a $24-million coal-cleaning plant at Meadow Creek, West Virginia."

* * * * *

"A fifth operating division, Beckett Tilford, Inc., located at Palm Beach International Airport, was acquired by Beckett Aviation in 1979, joining other units in Cleveland, Chicago, Pittsburgh and Youngstown."

"This full-service fixed-based operation incorporates a flight school and charter operation to compliment the traditional Beckett services."

* * * * *

"Beckett employes 400 people, of whom 100 are pilots, and operates fifty-two aircraft ranging from small two-place trainers through turboprops, helicopters and pure jet equipment, including intercontinental-type airliners."

* * * * *

In the reshuffling of the management team that transpired while the new headquarters was being transferred to Richmond, Jack Ford had evidently suffered the same fate of being passed over as he had been forced to inform me the year before. When he learned that he would not be included among the officers that were going to Richmond, he immediately went to his office, cleaned out his desk and left a brief note of resignation.

A very welcome note arrived for me from Cleveland in the middle of the year notifying me that I was to receive a twelve percent raise effective immediately. This was the biggest salary boost I had ever received at one time and I surmised that someone in the upper echelons must have come to my aid. This news really warmed the cockles of my heart because I hadn't had to fight for it and it would help me in retirement income.

Probably the most enjoyable side benefit of my job was the opportunity to make frequent visits to the Greenbrier hotel. This place has a way of making a profound impression on the wife. One time Carl Del Signore told me that he never found as fine accommodations and hospitality in all of his travels around the world. I have enjoyed having many friends of the staff there and feel like it is my second home. The management keeps making improvements that add to the comfort of the guests.

The clinic at the Greenbrier is located on the fifth floor of the West Virginia Building. The facilities are designed around pleasant surroundings and the personal treatment is excellent. The C&O first sent me to the clinic for a physical examination in 1956 and I have just completed my 37th visit. My doctor for his first 27 trips was Gene Morhouse and is now Thomas Mann. They seem to be completely cooperative and helpful for the patient.

☆

1980 - MERGER WITH FAMILY LINES

The merger of the Chessie System and Seaboard became effective on November 1 creating the CSX Corporation. The headquarters for the new company was established in Richmond, Virginia in the Federal Reserve Building. The first stockholder's meeting was held there and the information at the meeting was presented by Hays Watkins as President and Prime F. Osborn III as Chairman. Their joint presentation is excerpted as follows:

"The year 1980 was a remarkable and a rewarding one for the company. It saw the establishment of CSX, the passage of landmark railroad deregulation legislation, the beginning of a long-term steam coal export boom and strong earnings."

* * * * *

"The merger was clearly the highlight of 1980. It was approved by the Interstate Commerce Commission in late September, only 20 months after filing - a record for merger cases of this magnitude. On November 1, CSX began operations with headquarters in Richmond, Virginia. And it is quite a company! We have over 70,000 employees, nearly $5 billion in revenues and assets in excess of $7.5 billion."

"We are a large, diversified company with a prominent position in many fields. With the merger, CSX railroads became the largest rail system in the country. Our railroads offer single-system service in 22 states, the District of Columbia and the Province of Ontario, Canada. While the bulk of our revenues and assets are in the railroad industry, we also have vital enterprises and great strength in real estate, oil and gas, cable TV, aviation, resort hotel, publishing, data processing, and natural resources development activities."

* * * * *

"The two major rail units, Chessie System and The Family Lines, will continue to be run separately from Cleveland and Jacksonville, respectively. However, they are being operated under general policy guidance from CSX to maximize the service efficiencies and revenue benefits for the total enterprise."

* * * * *

"Earnings for 1980 were $281.6 million, or $7.13 per share, after on-time, merger-related accounting adjustments which reduced 1980 earnings by $20.4 million or 52 cents per share. These adjustments, resulting from the 'pooling of interests' accounting method, were adopted to conform the accounting principals and financial reporting of the former companies. If Chessie System, Inc., and Seaboard Coast Line Industries, Inc., had reported separately, each would have had another year of record earnings."

"Enactment of landmark railroad regulatory reform legislation, the Staggers Act, was another major development of the past year. This legislation makes far-reaching changes in century-old railroad regulatory concepts. While not total deregulation by any means, the Staggers Act does recognize the need for the rail industry to have more pricing flexibility."

* * * * *

"Another highlight of the year was export coal with dumpings of nearly 35 million tons, up 70 percent from 1979, making CSX rail group the leader in coal exports. Most of this increase came from steam coal, where we saw the beginning of a significant, long-term increase in demand from abroad. The sudden surge in demand placed a serious strain on existing port capacity, with the resulting highly-publicized ship delays at Hampton Roads and Baltimore. Fortunately, the current bottleneck promises to be a relatively short-term situation as a number of plans are already well under way to add new piers and expand the capacity of existing facilities."

* * * * *

A Consolidated earnings statement for the new company over the past three years was presented:

	1980	1979	1978
		(Thousands of Dollars)	
Operating Revenues			
Transportation	**$4,497,013**	$4,026,954	$3,472,464
Non-Transportation	**344,337**	326,926	219,061
	$4,841,350	$4,353,880	$3,691,525
Earnings for the Year	**$ 281,607**	$ 237,122	$ 131, 303
Earnings Per Share	**$7.13**	$6.12	$3.37
Total Assets	**$7,523,101**	$7,082,010	$6,739,921
Long-Term Debt	**$2,067,997**	$2,062,264	$2,235,653

The remainder of the annual report was devoted to a discussion of the development of the newly merged CSX Company.

"The first single-system unit coal train was operated in mid-November from a mine in West Virginia to an electric generating plant at Gainsville, Florida. With the new expedited service, a round trip is being made every 10 days."

"Unit grain trains are being handled in single-system service from the mid-west to elevators at Hampton Roads, Virginia, over a direct, combined rail route. The first train arrived November 30. Also, single-system piggyback service is being provided between Philadelphia and Jacksonville."

* * * * *

"This year additional merger plans will be implemented:

. . . A track connection at Deane, Kentucky, to permit the expedited handling of single-system trains, saving as much as 200 miles on some coal movements.

. . . Consolidation of multiple yard operations at East St. Louis, Illinois.

. . . Coordinated operations from Cincinnati into Osborn Yard at Louisville.

. . . Single-system operations between Winchester and Lexington, Kentucky, permitting abandonment of 10 miles of track and consolidation of yards at Lexington.

. . . Consolidation of piggyback operations in Cincinnati."

"The new car manufacturing facility at Raceland, Kentucky, is being used to build new coal hopper cars for the Family Lines. A total of 650 cars were delivered by year end at considerable savings."

* * * * *

"CSX Resources was established to provide centralized management of all real estate, oil, gas and timber properties."

"CSX Minerals was incorporated to provide unified management of all coal-land development activity and assist coal companies in acquiring mineable coal reserves and opening mines."

"CSX Beckett Aviation assumed operating responsibility for an additional fixed-based facility at Richmond, its tenth location."

"CSX expanded its cable television activities, completing several purchases in Florida."

* * * * *

"During the year, 177 new diesel-electric units were acquired. In addition, 46 older diesel locomotives were rebuilt at an engine facility in Tampa at less than half the price of new units - a savings of about $350,000 per locomotive."

"Over 2,800 new freight cars were acquired and more than 1,200 cars were reconstructed, even though total carloadings were down slightly for the year."

"Maintaining the railroads' physical plant to high standards is essential to the continuation of efficient operations. CSX rail units spent $535.4 million during the year on roadway, yards, and equipment - an all time high. Included were over 950 miles of smooth-riding, long-wearing, continuous welded rail; about five million cross ties, 10,685 miles of track surfacing and the addition of more than five million tons of ballast."

* * * * *

"While track work took place all across the widespread system, some areas received special attention. For example, a 42-mile segment of track between Atlanta and Cartersville, Georgia, which carries nearly 50 million tons annually and is one of the most heavily used mainlines in the nation, was extensively rebuilt. Included in the project was undercutting and ballast rock renewal of six miles of track, 39 switches and the reworking of 32 road crossings and seven sidings."

* * * * *

"Queensgate Terminal, the $71 million classification yard in Cincinnati, was partially opened for use during the year. The entire complex is scheduled for full operation by mid-1981."

"In Toledo, a $30 million iron ore pellet terminal is well underway, with completion set for this summer."

"Several large bridge projects were underway at year's end, including a new $7 million structure across the Tennessee River in Alabama, and a new $1 million bridge over Swift Creek in Virginia."

"The Biloxi Bay Bridge near Biloxi, Mississippi, was replaced at a cost of approximately $21 million."

* * * * *

"Major expansion projects completed were the Through Bulk Service terminals at Columbus, Ohio, and Detroit; piggyback terminals at Chicago and Lordstown, Ohio; an automobile loading facility near Cincinnati; double-tracking of the Big Sandy sub-division; a $4.5 million spur line to a new mine at Mannington, West Virginia; reactivation of Pier 15 at Newport News, and expansion and improvement of the coal terminal at Paintsville, Kentucky."

* * * * *

"Final plans were completed for development of 1,600 acres in suburban Loudon County, Virginia, near Washington, for a planned community of 4,000 single and multiple family residential units. A rezoning application is pending. The project will include office and retail facilities."

"Greenbrier Village, a project of the Old White Development Company, a wholly-owned CSX subsidiary, moved from the drawing boards to actual construction in 1980. The planned community, adjacent to The Greenbrier, will be one of the most prestigious and exclusive residential developments in the nation."

* * * * *

"CSX Resources participates, through joint ventures in Texas, Louisiana, Mississippi and new Mexico, as a working partner in lease acquisition, seismographic studies and the drilling of wells. Over the past three years, the value of reserves, as estimated by outside engineers, has increased five-fold from a cumulative investment program which has grown three-fold."

* * * * *

"CSX holds interest in 200 wells either currently producing or awaiting hook-up to pipelines."

* * * * *

"The new sawmill of Eastern Hardwoods, Inc., a CSX joint venture, reported a profit in its first full year of operations even though it was adversely affected by the recession, particularly in the housing industry,"

* * * * *

"In 1980, CSX Mineral Development Company was incorporated. When fully staffed it will assist companies in acquiring coal reserves and opening mines in Eastern Kentucky to be served by CSX rails."

* * * * *

The ascendance of Reagan to the Presidency gave the coal industry great hopes for a new era in the industry which soon turned to more gloom due to excess production capacity.

It was hard for me to realize that I would reach the accepted retirement age of 62 in the month of October. On my first visit to Cleveland in February, Jerry Gobrecht asked me if I was going to retire at 62. Some officers had gotten court decision allowing them to stay on until age 65. I could tell that there was no great desire on the part of management to keep me on any longer and one or two of my subordinates were showing signs of coveting my job. Therefore, I assured Jerry that I would depart at the end of my birth month of October. I planned to open a consulting office and felt that there was a chance I could make something of the idea.

Right near the first of October, I received a personal letter from John T. Collinson, President, advising me that I had been included on a new list of officers eligible for a new stock appreciation plan. I was awarded over a thousand points in the plan which was to offset a similar plan on the Family Lines.

With a few simple calculations, I could see that any increase in the price of our stock over the next four years would provide me the opportunity to convert my points into stock and cash. It was therefore obvious that I could make a lot of money. When this dawned on me, I felt an exhilaration like winning a lottery. For several days, I walked around like treading on a cloud. This plan really changes my retirement income from meager to almost satisfactory. I had never earned in salary what I thought I deserved and this windfall tended to assuage my bruised feelings.

CSX CORPORATION OFFICERS

Prime F. Osborn, III . . . . *Chairman of the Board*
Hays T. Watkins . . . . *President*
Robert L. Hintz . . . . *Senior Vice President-Finance*
John W. Snow . . . . *Senior Vice President-Corporate Services*
Josiah A. Stanley, Jr. . . . . *Senior Vice President-Audit*
Leonard G. Anderson . . . . *Vice President and Treasurer*
Edwin E. Edel . . . . *Vice President-Corporate Communications*
Carl C. Hawk . . . . *Vice President and Corporate Secretary*
James T. Lyons . . . . *Vice President-Taxes*
Gerald L. Nichols . . . . *Vice President and Executive Assistant*
Woodruff M. Price . . . . *Vice President-Government Relations,*
Washington, D.C.
Garth E. Griffith . . . . *General Counsel*

CHESSIE SYSTEM RAILROADS

John T. Collinson .... *President and Chief Executive Officer*

Roland W. Donnem .... *Senior Vice President-Law*

Norman G. Halpern .... *Senior Vice President*

John S. Lanahan .... *Senior Vice President-Commercial*

James B. McCahey, Jr. .... *Senior Vice President-Merchandise Traffic*

Alvin R. Carpenter .... *Vice President-Coordinations*

Don S. Garda .... *Vice President-Labor Relations*

Jerry E. Gobrecht .... *Vice President-Coal Traffic*

Paul R. Goodwin .... *Vice President-Finance*

Charles J. Henry, Jr. .... *Vice President-Merchandise Pricing*

William F. Howes, Jr. .... *Vice President-Casualty Prevention*

Thomas A. Keefe ..... *Vice President-Western Sales, Detroit, Michigan*

Jerome D. Krassenstein .... *Vice President-Marketing Services*

Kenneth C. Morriss .... *Vice President-Engineering and Maintenance*

William L. Ollerhead ... *Vice President-Eastern Sales, Baltimore, Maryland*

Richard G. Rayburn .... *Vice President-Transportation*

Patricia J. Hunady .... *Corporate Secretary*

Lester C. Roig, Jr. .... *Assistant Vice President and Treasurer*

Bruce G. Lawler ...... *Asst. Vice President and Comptroller, Baltimore*

Milton B. Dolinger .... *Director Public Relations and Adv., Asst. Vice President*

Richard B. Allen .... *General Counsel*

John H. Gobel .... *General Counsel*

William P. Coliton .... *President, Western Maryland Railway, Baltimore, Maryland*

Albert W. Dudley .... *President, Chicago So. Shore & So. Bend RR, Michigan City, Indiana*

THE FAMILY LINES RAIL SYSTEM

A. Paul Funhouser .... *President and Chief Executive Officer*
Welborn E. Alexander, Jr. ..... *Exec. Vice President-Sales and Marketing*
Philip M. Lanier .... *Executive Vice President-Law and Public Affairs*
Richard D. Spence .... *Exec. Vice President-Operations*
Alex J. Mandl .... *Senior Vice President-Finance*
Charles E. Mervine, Jr. .... *Vice President-Personnel and Labor Reltaions*
Richard D. Sanborn .... *Senior Vice President-Administration*
James D. Bozard .... *Vice President-Sales*
Jack M. Cooper .... *Vice President-Management Information Services*
Aubrey M. Daniel .... *Vice President-Phosphates and Chemicals*
Kenneth C. Dufford ...... *Vice President-Operations, L&N, Louisville, KY*
C. Hayden Edwards .... *Vice President-Secretary*
James W. Hoeland .... *Vice President-Marketing*
Philip L. Hooper .... *Resident Vice President, L&N, Nashville, TN*
William I. Johnson .... *Vice President-Director of Purchasing, Louisville*
A. Crane Jones, Jr. .... *Vice President-Operations, SCL*
Chester R. Lapeza .... *Vice President-Risk Management*
Charles D. Leddon .... *Vice President Staff-System Operations*
J. Ross LeGrand .... *Vice President-Industrial Development*
Joseph L. Lenihan ... *Vice President and General Counsel, L&N, Louisville*
Donald L. Morris .... *Vice President and Comptroller*
John E. Nall .... *Vice President-Coal Traffic, Louisville*
Hershel W. Parmer .... *Resident Vice President, Atlanta, GA*
Harold L. Snyder .... *Vice President -Treasurer*
John W. Weldon .... *Vice President-Law, SCL*
John W. Thomas .... *General Manager, Clinchfield RR Co., Erwin, TN*
M. S. Jones .... *General Manager, Georgia Railroad, Atlanta*

The Chessie System has been most fortunate in having a leader as wise and capable as Hays Watkins. Much of the success of the company has been the result of his imagination and steady hand on the tiller. Although I never reported to him directly, I had many opportunities to meet and talk to him at business meetings and on social occasions. He always impressed me as a real gentleman who was comfortable in control of the situation and conducted himself with confidence and cordiality. It is very doubtful whether the CSX merger would have come about without his insight and mediating skills.

It was a rare coincidence that the CSX merger became effective the same month of 1980 in which I retired.

☆

POSTSCRIPT

After a thirteen year interlude since my retirement, I can truthfully say that I feel personal satisfaction in the accomplishments of my career and confess that most of my experience worked out for the best. I thoroughly enjoyed the challenge of my job and the contributions that our company was making in the economic development of our country.

I do not regret missing out on the tremendous changes that have taken place in the company since I retired in 1980. After pleading with management for several years to obtain a computer for our office so we could keep up with changes all around us, the computer was in the local warehouse ready to be installed when I left. I can't say that I am sorry that I didn't have to put it to use.

Several years ago the coal lands were sold to a group of investors in Texas, eliminating one of the sources of outside income for the company. Pete Carpenter was obviously on the way up in management in 1980 and is now running the railroad.

One year after I retired I learned that Conrail was going to build a branch line to serve a new mine on the Lewis Land property. This meant that the project area where I started my career in 1946 and spent much time promoting and defending throughout my career was finally being lost to the opposition. At least I can say that it didn't happen on my watch.

The operating headquarters of the company has been moved south to Jacksonville which is a long way from the coal fields. The drab blue and gray colors adapted by the new company were depressing compared to the bright engines of the Chessie System. Recent changes of adding bright yellow front on the CSX engines has improved their appearance.

From the recent improvements in the stock prices of the company, it is obvious that many kinks in the new organization have finally been worked out. Since we have a definite financial interest in the success of the company, all of the retirees are hoping for the best. The Cando Club, made up of former employees, is still very active in Huntington. The C&O Credit Union is still active although the locomotive shops is the only major company activity in the city besides a division headquarters.

A very complimentary article about the company has just appeared in the June 7 issues of *Forbes Magazine* as it is being written:

"Full steam ahead."

"Forbes went out on a limb in proclaiming that railroad laggard CSX Corp. had rounded the bend and was picking up speed under the new Chief Executive John W. Snow, a former federal highway official. For a while it looked as if the limb were sawed off: CSX's stock proceeded to slump to the mid-20's from the mid-30's.

"But since then the $8.7 billion (revenues) giant has powered ahead. As Snow has streamlined operations and negotiated down the labor costs, CSX's stock surged to 80 early this year, before correcting to a recent 71-5/8. That made CSX the best-performing company in a very hot transportation group over that 35-month period."

"CSX has yet to realize all of the cost savings from the improved labor agreement it has hammered out. The railroad's expenses have been chopped to an estimated 81% of revenues from 88% in 1990, but Snow aims to get them below 80%. Every percentage point is worth roughly 50 cents per share."

While playing golf last week, I met a young man by the name of Larry Roberts who is a trainmaster on the Huntington District. He is a handsome, polite young man with a positive attitude which type of employee bodes well for the future of CSX.

From my own perspective, I see a continuing bright future for the world's largest coal-hauling railroad and wish it great prosperity and success.

☆